# SERVICE-HANDBOOK ROLLS-ROYCE SILVER DAWN, SILVER WRAITH, PHANTOM IV AND BENTLEY MK. VI, R-TYPE

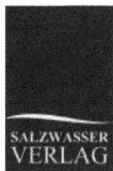

SALZWASSER
VERLAG

www.salzwasserverlag.de

**SERVICE-HANDBOOK ROLLS-ROYCE SILVER DAWN, SILVER WRAITH, PHANTOM IV AND BENTLEY MK. VI, R-TYPE**

1. Auflage 2009 | ISBN: 9783941842441

Salzwasser-Verlag (www.salzwasserverlag.de) ist ein Imprint der Europäischer Hochschulverlag GmbH & Co KG, Bremen. (www.eh-verlag.de). Alle Rechte vorbehalten.

Die Deutsche Bibliothek verzeichnet diesen Titel in der Deutschen Nationalbibliografie.

# SERVICE
# HANDBOOK

SILVER WRAITH — SILVER DAWN — BENTLEY MK. VI.
R. TYPE BENTLEY — PHANTOM IV.

# SERVICE HANDBOOK

## SECTION INDEX

(069CPM-2234F)
DEPT 76 A-4
REFER 8-7-51

SERVICE HANDBOOK

SILVER WRAITH — SILVER DAWN — BENTLEY MK. VI.
R. TYPE BENTLEY — PHANTOM IV.

# SECTION
# A
# GENERAL INFORMATION

## SILVER WRAITH — SILVER DAWN — BENTLEY MK. VI.
## R. TYPE BENTLEY — PHANTOM IV.

### S P E C I F I C A T I O N

#### SILVER WRAITH, SILVER DAWN, BENTLEY, PHANTOM IV

**ENGINE AND CHASSIS NUMBERS:**

- The engine number is stamped either on the front left-hand crankcase lifting lug, or on a boss on the crankcase immediately above.

The chassis number will be found on the identification plate, fixed to the front of the dashboard, and also stamped on the left-hand side frame member just in front of the dashboard, under the bonnet.

**ENGINE:**

Type

- Six cylinders, in line, with overhead inlet and side exhaust valves.

#### SILVER WRAITH

**WTA-1 - WME-67.**

| | |
|---|---|
| Bore | - 3.500" (88.9 mm) |
| Stroke | - 4.500" (114.3 mm) |
| Cubic capacity | - 260 cu.ins. (4,257 c.c) |

**WOF-1 - and onwards.**

| | |
|---|---|
| Bore | - 3.625" (92 mm) |
| Stroke | - 4.500" (114.3 mm) |
| Cubic capacity | - 279 cu.ins. (4,566 c.c) |

**Compression Ratio.**

| | | |
|---|---|---|
| WTA-1 | - WSG-101 | - 6.4:1 (Cylinder head RE.10429) |
| WVH-1 | - and onwards | - 6.4:1 (Cylinder head RE.10429) |
| | | - 6.75:1 (Cylinder head RE.13451) |

#### SILVER DAWN

**SBA-2 - SDB-200.**

| | |
|---|---|
| Bore | - 3.500" (88.9 mm) |
| Stroke | - 4.500" (114.3 mm) |
| Cubic capacity | - 260 cu.ins. (4,257 c.c) |

**LSFC-2 - and onwards.**

| | |
|---|---|
| Bore | - 3.625" (92 mm) |
| Stroke | - 4.500" (114.3 mm) |
| Cubic capacity | - 279 cu.ins. (4,566 c.c) |

# SERVICE HANDBOOK

Compression Ratio.

SBA-2 - SHD-60    - 6.4:1 (Cylinder head RE.10429)
SKE-2 - and onwards    - 6.4:1 (Cylinder head RE.10429)
   - 6.75:1 (Cylinder head RE.19451)

#### BENTLEY

B-2-AK - B-401-LH

Bore    - 3.500" (88.9 mm)
Stroke    - 4.500" (114.3 mm)
Cubic capacity    - 260 cu.ins. (4,257 c.c)

B-2-HD - and onwards.

Bore    - 3.625" (92 mm)
Stroke    - 4.500" (114.3 mm)
Cubic capacity    - 279 cu.ins. (4,566 c.c)

Compression Ratio.

B-2-AK - B-301-PU    - 6.4:1 (Cylinder head RE.10429)
B-2-RT - and onwards    - 6.4:1 (Cylinder head RE.10429)
   - 6.75:1 (Cylinder head RE.19451)
BC-1A - BC-57B    - 7.27:1 (Cylinder head RE.16876)
   - 7.20:1 (Cylinder head RE.19451)

#### PHANTOM IV

4-AF-2 - and onwards.

Bore    - 3.500" (88.9 mm)
Stroke    - 4.500" (114.3 mm)
Cubic capacity    - 346 cu.ins. (5,675 c.c)

Compression Ratio.    - 6.4:1

Suspension.    - The engine and gearbox are of unit construction, the engine being flexibly mounted on rubber at two points.

Torsional rigidity is controlled by a torque arm, fixed to the rear of the gearbox, bearing on two rubbers. Fore and aft location is obtained by a tie-bar.

CYLINDER BLOCK:

Type    - Monobloc casting, integral with crankcase.

Material    - Cast iron.

SILVER WRAITH — Commencing at chassis WOB-31, short "Bricrome" inserts ($2\frac{1}{4}$" long) are pressed into top of cylinders. Commencing at chassis WAB-14, phosphor-bronze replace cast iron exhaust valve guides.

SILVER DAWN — "Bricrome" inserts ($2\frac{1}{4}$" long) are pressed into top of cylinders. Phosphor bronze exhaust guides.

BENTLEY — Commencing at chassis B-144-DA, short "Bricrome" inserts ($2\frac{1}{4}$" long) are pressed into top of cylinders. Commencing at chassis B-26-CF, phosphor bronze replace cast iron exhaust valve guides.

PHANTOM IV — "Bricrome" inserts ($2\frac{1}{4}$" long) are pressed into top of cylinders. Phosphor bronze exhaust guides.

CYLINDER HEAD:

Type — Detachable.

Material — Aluminium alloy, with nickel chrome steel inlet valve seats and cast iron guides.

CRANKSHAFT:

Material — Nitrided chrome molybdenum steel. Fully machined and balanced.

Number of journals — Silver Wraith - Seven.
Silver Dawn - Seven.
Bentley - Seven.
Phantom IV - Nine.

Balance weights — Silver Wraith, Silver Dawn and Bentley detachable.

Crankshaft vibration damper — Silver Wraith, Silver Dawn and Bentley, internal, combined spring drive and friction type damper.
Phantom IV - rubber tuned harmonic balancer.

MAIN BEARINGS:

Number of — Seven - Silver Wraith, Silver Dawn and Bentley.
Nine - Phantom IV.

Type — Copper, lead-indium lined thin steel shells with "pre-sized" bores to suit diameter of crankshaft journals.

CONNECTING RODS:

| | |
|---|---|
| Type | - 'H' section. Fully machined and balanced. |
| Material | - Chrome Molybdenum steel. |
| Big-end bearings, type | - Copper, lead-indium lined thin steel shells with "pre-sized" bores to suit diameter of crankpins. |
| Gudgeon pin bush | - Pressed in rod. |

PISTONS:

| | |
|---|---|
| Material | - Aluminium alloy, tin plated. |
| Type | - Early series Silver Wraith and Bentley fitted with solid skirt pistons. Later series and Silver Dawn and Phantom IV, split skirt pistons used. |
| Number of rings | - Two compression and one slotted oil scraper. |

CAMSHAFT:

| | |
|---|---|
| Material | - Case hardening nickel steel. |
| Number of journals | - Silver Wraith, Silver Dawn and Bentley, four.  Phantom IV, six. |
| Bearings | - Babbit lined steel shells. |
| Thrust taken | - Front. |
| Drive | - Helical toothed gear. |
|     SILVER WRAITH | - Chassis Nos. WTA-1 to WAD-86, fabric type driving gear. Chassis WHD-87 and onwards, aluminium driving gear. |
|     SILVER DAWN | - Chassis Nos. SBA-2 to SCA-21, fabric type driving gear. Chassis Nos. LSCA-23 and onwards, aluminium driving gear. |
|     BENTLEY | - Chassis Nos. B-2-AK to B-401-GT, fabric type driving gear. Chassis Nos. B-2-HR and onwards, aluminium driving gear. |
|     PHANTOM IV | - Aluminium driving gear. |

# SERVICE HANDBOOK

### SILVER WRAITH — SILVER DAWN — BENTLEY MK. VI.
### R. TYPE BENTLEY — PHANTOM IV.

VALVE GEAR:

General
- Dual inlet valve springs. One overhead inlet valve per cylinder operated by push rods. Gland packings on inlet valve stems to control air leakage and lubrication. One side exhaust valve per cylinder. Exhaust valves have "Stellited" facings; a special heat resisting material.

Valve tappets
- Barrel type, flat face.

LUBRICATION SYSTEM:

General
- High pressure feed to crankshaft, connecting rod and camshaft bearings and the distributor drive skew gear. Dual oil relief valve providing a positive low pressure oil supply to engine gears and to the hollow valve rocker shaft from which valve rockers, push rods, tappets and cams are lubricated.

Type
- Pressure throughout.

High pressure supply
- 25 lbs/sq.in. (approx).

Low pressure supply
- 5 lbs/sq.in. (approx).

Oil pump
- Spur gear type with floating intake strainer.

Oil pressure relief valve unit
- Dual type, controlling both high and low pressure feeds.

Oil filter, types

    SILVER WRAITH
- A - E series chassis, 'General' By-Pass Oil Filter. F series and onwards, 'Vokes' Full Flow Filter.

    SILVER DAWN
- A and B series chassis, 'General' By-Pass Oil Filter. C series and onwards, 'Vokes' Full Flow Filter.

    BENTLEY
- A - L series chassis, 'General' By-Pass Oil Filter. M series and onwards, 'Vokes' Full Flow Filter.

    PHANTOM IV
- A series chassis, 'General' By-Pass Oil Filter. B series and onwards, 'Vokes' Full Flow Filter.

SILVER WRAITH — SILVER DAWN — BENTLEY MK. VI.
R. TYPE BENTLEY — PHANTOM IV.

FUEL SYSTEM:

Carburetter Types

Make and Type

SILVER WRAITH
- R.H. and L.H. drive chassis,
WTA-1 to WSG-5 - Stromberg Type
AAV-26M, Dual Downdraught pattern.
WSG-7 and onwards - Zenith Type
LDVC.42, Single Downdraught pattern.

SILVER DAWN
- R.H. and L.H. drive chassis, SBA-2
to LSFC-100 - Stromberg Type AAV-26M,
Dual Downdraught pattern. LSFC-102
and onwards - Zenith Type LDVC.42,
Single Downdraught pattern.

BENTLEY
- Twin S.U. Type H.4 ($1\frac{1}{2}$" choke).
Chassis B-2-AK to B-81-HP, R.H.
drive cars only.
Twin S.U. Type H.6. ($1\frac{1}{4}$" choke).
Chassis B-83-HP to 'R' series, R.H.
drive chassis.
Stromberg Type AAV-26M, Dual Down-
draught pattern, L.H. drive chassis
to 'R' series.
Twin S.U. Type, H.6. Fully Automatic,
'R' Type and onwards.
Continental Series BC-2-LA to
BC-6-LA, Non-Automatic.
BC-8-LA, Fully Automatic.

PHANTOM IV
- Stromberg Type AAV-26M, Dual Down-
draught pattern.

Air Cleaner
- Mesh or Oil Bath.

Fuel Pumps
- S.U. Twin electric, Type L,
Phantom IV - Heavy duty type.

Fuel Tank capacity
- 18 gallons (Imperial).

Fuel Strainers
- Main fuel strainer mounted on frame
cross-member in front of fuel tank.
In addition, a small gauze strainer
is arranged on the carburetter/s.

Fuel Gauge
- Electric. Registers when the master
and ignition switches are "on".

COOLING SYSTEM:

Cooling system capacity
- Silver Wraith, Silver Dawn, Bentley
- 4 gallons (Imperial).
Phantom IV - 5 gallons (Imperial).

Type
- Pressure.

Pump
- Centrifugal.

# SERVICE HANDBOOK

### SILVER WRAITH — SILVER DAWN — BENTLEY MK. VI.
### R. TYPE BENTLEY — PHANTOM IV.

COOLING SYSTEM:  (Cont'd).

| | |
|---|---|
| Fan | - Five blade. |
| Fan diameter | - 16" dia. fan - Silver Wraith WTA-1 to WHD-75, Silver Dawn SBA-2 to LSCA-7, Bentley B-2-AK to B-211-GT. Phantom IV - all. 17¼" dia. fan - Silver Wraith WHD-77 and onwards. Silver Dawn LSCA-9 and onwards. Bentley B-213-GT and onwards. Fans are not interchangeable. |
| Pump and fan drive. | - By adjustable Vee belt. |
| Radiator matrix, type | - Corrugated. |
| Radiator shutters | - Silver Wraith - Thermostatic operated. Silver Dawn  - Fixed. Bentley        - Fixed. Phantom IV   - Thermostatic operated. |
| Coolant temperature control | - Thermostatically controlled by a by-pass thermostat, allowing a minimum running coolant temperature of 80°C. |
| Temperature indicator | - On instrument panel. |
| Coolant | - The cooling system is filled with a 25% mixture of inhibited ethylene glycol and water (anti-freeze) before the car leaves the factory. |

EXHAUST SYSTEM:

| | |
|---|---|
| General | - The exhaust system comprises four main parts, the front exhaust pipe, two separate silencers and the tail pipe. The front silencer which is of oval section has one baffle through which pass one short and two long perforated tubes. The rear silencer is circular and consists of a resonator chamber followed by three cylindrical co-axial baffles. The exhaust line runs down the left-hand side of the chassis, the tail pipe being carried over the rear axle. The exhaust system is flexibly mounted and is thus at liberty to float in unison with the engine unit. Commencing at Bentley B-2-MD, a twin exhaust system is fitted (to R.H. drive chassis only), consisting of one front silencer and one rear silencer in each branch. Each silencer is of oval section, having a centre baffle plate through which pass three long perforated tubes. |

- A.7 -

CLUTCH:

| | |
|---|---|
| Make | - Borg and Beck (Semi-centrifugal). Single dry plate. |
| Size and type | |
| SILVER WRAITH | - 11" 'Heavy', except WTA-1 to WTA-55 - 10" 'Long' type. |
| SILVER DAWN | - 10" 'Long' type, SBA-2 to SCA-25. 11" 'Light' type, SCA-27 to SDB-74. 11" 'Heavy' type, SDB-76 and onwards. |
| BENTLEY | - 10" 'Long' type, B-2-AK to B-402-GT. 11" 'Light' type, B-2-HR to B-298-LJ. 11" 'Heavy' type, B-300-LJ and onwards. |
| PHANTOM IV | - 11" 'Heavy' type. |
| Facing material | - Mintex H-14. |
| Clutch Pressure springs | - Nine. The original 10" clutch was equiped with orange coloured pressure springs. These have been superseded by red coloured springs. The 11" 'Light' and the 11" 'Heavy' type clutches are both equiped with orange coloured springs, except Bentley Continental which has six red and three orange. Phantom IV - red springs. |
| Damper springs | - The driven plate damper springs are red coloured for all clutches, except Phantom IV - these are orange. |
| Clutch release bearing | - Ball bearing (single row). Lubricated from chassis Luvax system. |

GEARBOX: (Synchromesh)

| | |
|---|---|
| Number of speeds | - Four speeds and reverse, with positive interlock selector mechanism. Synchromesh on 2nd, 3rd and 4th speed gears. |
| Gearbox ratios | - 1st speed - 2.98:1 2nd speed - 2.02:1 3rd speed - 1.34:1 4th speed - Direct Reverse  - 3.15:1 |
| Oil capacity | - 6 Pints (Imperial) 7.2 Pints (U.S.A) 3.4 litres. |

GEARBOX:  (Automatic)

General
- Optional on later series left-hand drive cars. A manual control lever on the steering column allows selection of desired ranges to suit operating conditions.
Three forward speed ranges are provided also neutral and reverse. Control lever quadrant marked N. 4. 3. 2. R.

Gearbox ratios
- 1st speed - 3.82:1
2nd speed - 2.63:1
3rd speed - 1.45:1
4th speed - Direct
Reverse   - 4.30:1

Oil capacity
- 20 Pints (Imperial) 24 pints (U.S.A) 11.36 litres.

PROPELLOR SHAFT:

Type
- Divided open type having three needle roller bearing universal joints; one near the centre and one at each end. The shaft is supported in the centre by a flexible mounted ball bearing.

REAR AXLE:

Type
- Semi-floating.

Final drive
- .650" off-set Hypoid bevel gears.

Oil capacity
- Silver Wraith, Silver Dawn and Bentley - 1¾ pints (Imperial).
Phantom IV - 3 pints (Imperial.)

Pinion Thrust Bearing.
- As from Silver Wraith Chassis WVH-72, Silver Dawn Chassis SLE-11 and approx. the middle of Bentley SR series, a pair of Tinker roller bearings replaces the double thrust bearing. An internal flange separates the two bearings. These are only fitted to chassis with 11:41 or 12:31 ratio pinions.

BRAKES:

General
- Hydraulic operation on front wheels, mechanical operation on rear wheels assisted by a mechanically driven servo motor. The foot pedal operates the rear brakes and also engages the servo. The servo which operates the front brakes, through the medium of a balance lever and a hydraulic master cylinder, at same time augments the direct application of the rear brakes.

BRAKES:   (Cont'd)

General, continued
- With the leverages provided, about 55% of the total braking is imposed on the front wheels to counteract the greater weight thrown upon the front wheels when braking.  The hand brake operates directly on the rear brakes only.

Brake shoe linings
- Ferodo VG-91.

Friction lining area (4 brakes)
- Silver Wraith, Silver Dawn, Bentley - 186.6 sq.in.
Phantom IV - 272 sq.in.

Hand brake lever
- "Scuttle" type.

SERVO MOTOR:

General
- Dry single disc brake type, driven from the gearbox at approx. 1/10th propellor shaft speed, and is actuated by means of axial cams whose levers are interposed in the rod from the foot brake pedal to the rear brakes intermediate shaft.  The servo motor operates equally well for forward or reverse motion of the car.
Provision is made for external adjustment of the servo.

Servo motor lining
- Ferodo VG-91.

FRONT HUBS:

General
- Two single row ball bearings. Commencing Silver Wraith chassis WME-1, Silver Dawn chassis SCA-1, Bentley chassis B-1-GT, two taper roller bearings replace the ball bearings.

WHEELS AND TYRES:

Wheels, type
- Bolted on pressed steel wheels with balance weights and covering discs.

Rim-wheel and tyre

    SILVER WRAITH
- Well-base rims, 5.00" x 17.00" or 6.50" x 16.00".
Dunlop Fort 'C' tyres, 6.50" x 17.00" or 7.50" x 16.00" may be fitted to standard wheel base chassis.  On cars for U.S.A., Canada, Australia and Egypt, 7.50" x 16.00" tyres are fitted.  Also to all long wheel base chassis.

WHEELS AND TYRES:   (Cont'd).

| | |
|---|---|
| SILVER DAWN | - Well-base rims, 5.00" x 16.00". Dunlop Fort 'C' tyres 6.50" x 16.00". |
| BENTLEY | - Well-base rims, 5.00" x 16.00" India Super Silent Rayon tyres 6.50" x 16.0. except Continental chassis which are India Speed Special, 6.50" x 16.00". |
| PHANTOM IV | - Semi-drop centre rims, 5.00" x 17.00" Dunlop Fort 'C' tyres 8.00" x 17.00". |

STEERING:

| | |
|---|---|
| General | - The movements of the steering wheel are transmitted by the pendulum lever and the side steering tube to a centr. steering lever/s situated on the rear of the front cross member of the fram. and then by the two cross steering tubes to each pivot axle. |
| Steering unit, type | - Cam and roller. Not irreversible. |
| Ratio | - 15.25:1, except Silver Wraith from chassis WVH-43, 18.7:1. |
| Drive | - Right-hand or left-hand. |
| Steering wheel diameter | - Silver Wraith, Silver Dawn, Bentley, 18". Phantom IV, 18.75". |
| Oil capacity of box | - 1¼ pints (Imperial). |

SUSPENSION:

| | |
|---|---|
| Front | - Independent front wheel suspension by means of coil springs controlled by hydraulic dampers. Side sway is checked by an anti-roll torsion bar. |
| Rear | - Semi-elliptic leaf springs controlled by hydraulic shock dampers. Later model Silver Dawn and Bentley, incorporate spring leaves that are longer, wider and reduced in number. The rear shackle brackets have been turned through 90°, the shackle eyes being above instead of below the side member. |
| Front shock dampers | - Hydraulic double acting. |
| | The shock damper consists of two pistons operating in oil filled cylinders. |

SUSPENSION:   (Cont'd).

Front shock dampers   (Cont'd).
- The oil is displaced from one cylinder to the other through drilled passages, the degree of damping being controlled by spring loaded valves.

Rear shock dampers
- Controllable hydraulic double acting.

The shock damper consists of a piston assembly operating in an oil filled cylinder; the oil being displaced from one end of the cylinder to the other, past spring-loaded valves. The loading of these valves and hence the degree of damping, is controllable through the "Ride Control" lever by means of a small gear type oil pump carried in a casing bolted to the gearbox, which maintains a pressure of oil in a system of piping coupled to each rear damper. This pressure is variable and is controlled through a relief valve operated by the hand lever; pressure being applied to the damper through the medium of a metallic bellows which isolates the pressure oil from the damper oil.

On later models the single forked link between the shock damper-arm and the spring plate has been replaced by two parallel links. A separate bracket for the lower Silentbloc bush is no longer provided, this now being integral with the spring plate.

CHASSIS LUBRICATION:

System
- Luvax Bijur centralised chassis lubrication system supplied by a pedal operated oil pump mounted on the dashboard.

Capacity, oil pump
- 2 pints (Imperial).

CHASSIS FRAME:

Type
- Both front and rear portions have box section side members and rigidity is further increased by a special cruciform stiffening member.

JACKING SYSTEM

General
- A Dunlop "Triple Screw" jack is provided which operates on slides fitted under each side of the body sill.

SILVER WRAITH — SILVER DAWN — BENTLEY MK. VI.

R. TYPE BENTLEY — PHANTOM IV.

JACKING SYSTEM:  (Cont'd).

General  (Cont'd).

~ at about the centre of chassis, also special bracket fitted to front members on certain cars.
Early models were fitted with Smith's "Bevelift" jack.

BATTERY:

Make and type

- Either, P & R Lagenite, 9 HZP9-S, or Exide, 6 MXP9-L.

Voltage

- 12 volts.

Capacity

- 55 ampere/hour at 20 hour rate.

Earth

- Positive side of battery to chassis frame.

IGNITION DISTRIBUTOR:

Make and type

- Delco-Remy.  Twin contact breaker type with synchronisable contact breaker arms.

Rotation

- Clockwise.

Advance mechanism, type

- Automatic (centrifugal governor).

Firing order

- Silver Wraith, Silver Dawn and Bentley, 1, 4, 2, 6, 3, 5, No.1 being the front cylinder.
Phantom IV, 1, 6, 2, 5, 8, 3, 7, 4, No.1 being the front cylinder.

IGNITION COIL:

Make and type

- Lucas B.12.

SPARKING PLUGS:

Make and type

- Champion type N.8 or Lodge type CLN.30, except Phantom IV and Bentley Continental, Champion type N.8 only.

DYNAMO:

Make and type

- Lucas R.A.5 (5") except early chassis which had Lucas C.45 PV.(4½).

Maximum output

- 26 amperes at 13.5 volts.

Drive

- By adjustable Vee bolt.

Voltage regulator and cut-out

- Temperature compensated type.

SILVER WRAITH — SILVER DAWN — BENTLEY MK. VI.

R. TYPE BENTLEY — PHANTOM IV.

STARTER MOTOR:

Make and type
- Lucas M-45G 12v with built in planitary reduction gear (1.9:1 ratio) and Bijur type drive incorporating a friction clutch.

Cranking speed
- 80 to 160 engine R.P.M. (under normal temperate climatic conditions).

Rotation
- Clockwise.

Pinion to flywheel ratio
- 14/118.

HORNS:

Make and type
- Lucas, either model WT.29, WT.614 or HF.1748.

DIRECTION INDICATORS:
- Lucas, arm type, model SF.34.N on early series, later model SF.80. Flashing light indicator fitted to L.H. drive export cars only.

WINDSCREEN WIPERS:
- Early chassis, Houdaille "Berkshire" type.
Later, Lucas Rack type.
Later series, Lucas two speed type, the motor has a thermostatic controlled cut-out which operates if wiper is overloaded. Cut-out switch will reset if "Parked" for 15 minutes.

HEADLAMPS:

Make and type

      SILVER DAWN and BENTLEY
- Lucas Mark I.

General
- A small red warning light, mounted in the speedometer, is illuminated whenever the headlamps are on the "Driving Beam" (Full on).
Provision is made so that the system can be altered to double filament dipping to the right in both headlamps for use in countries where the car is driven on the right-hand side of the road by:-
  a) Fitting Lucas 303 or Osram OS 516R right-hand dip double filament prefocus bulbs to both headlamps.
  b) Connecting together the two pink wires in the right-hand wing valance junction box.
Commencing at Bentley Chassis No. B-169-NY and Silver Dawn Chassis No. LSHD-56, Lucas Mark II Headlamps are fitted.

- A.14 -

### SILVER WRAITH — SILVER DAWN — BENTLEY MK. VI.
### R. TYPE BENTLEY — PHANTOM IV.

HEADLAMPS:   (Cont'd).

General, (Cont'd)

- These are similar to the Mark I in operation, but differ slightly in design. The reflector unit is secured to the lamp body by three spring loaded screws and these are operated to give the correct alignment. Certain Mark II lamps may be equiped with a fuse unit fitted to the lamp shell; the fuses are rated at 15 amps.

#### SILVER WRAITH and PHANTOM IV

Make and type

- Lucas, "R.100".

General

- The headlamps are controlled by two switches, the master switch on the switchbox and the foot switch for beam selection. A small red warning light, mounted in the speedometer, is illuminated whenever the headlamps are on the driving beam (full on). These headlamps are of the dipping reflector type, the foot switch normally operates to dip the left-hand headlamp and extinguish the right-hand headlamp. Change-over switch connections in the headlamps allow for this procedure to be reversed if the car is used where the rule of the road is to drive on the right-hand side.

FOG LAMP:

- One centre lamp fitted to Silver Wraith and Bentley. Two lamps fitted to Silver Dawn and Phantom IV.

FUSE BOX:

- Main fuse carried in separate box on Phantom IV and later models Silver Wraith, Silver Dawn and Bentley. Circuit fuses one strand of No.32 S.W.G. copper wire. Main fuse, three strands.

DEMISTER AND DEFROSTER:

General

- The normal system consists of directing pre-heated air through suitable vents in the capping rail on to the windscreen, assisted by a blower motor.
An electrically heated element was incorporated on the driver's side on early cars.
Later models incorporate a hot and cold air demisting system.

SERVICE HANDBOOK

SILVER WRAITH — SILVER DAWN — BENTLEY MK. VI.

R. TYPE BENTLEY — PHANTOM IV.


CAR HEATER:

General
- A hot water heater is normal equipment, under the front passenger's seat. Extra heaters can be fitted as required.

WINDSCREEN WASHER:

General
- Later models are fitted with the Trico Vacuum operated windscreen washer, the pump and reservoir being installed on the front of the dashboard and operated by a push button on the facia board.

COACHWORK:

SILVER WRAITH

Dimensions (Short Wheelbase chassis) -

| | | |
|---|---|---|
| Wheelbase | - 10' 7" | (322.58 cms) |
| Track, front (with 6.50" x 17.00" tyres)- | 56. 9" | (144.53 cms) |
| Track, front (with 7.50" x 16.00" tyres)- | 57. 9" | (147.06 cms) |
| Track, rear (with 6.50" x 17.00" tyres)- | 59. 8" | (151.9 cms) |
| Track, rear (with 7.50" x 16.00" tyres)- | 60. 8" | (154.43 cms) |
| Overall length, including bumpers (approx). | - 17' 6" | (533.4 cms). |
| Overall width, over wings (approx). | - 6' 3" | (190.5 cms) front and rear. |
| Overall height, unladen (approx). | - 5' 8" | (172.7 cms) |
| Turning circle diameter (to outside edge of tyre) | - 43' 5" | (1,323.34 cms) R.H. and L.H. locks. |

Dimensions (Long wheelbase chassis) -

| | | |
|---|---|---|
| Wheelbase | - 11' 1" | (337.82 cms) |
| Track, front (with 7.50" x 16.00" tyres)- | 57. 9" | (147.06 cms) |
| Track, rear (with 7.50" x 16.00" tyres)- | 64" | (162.56 cms) |
| Overall length, including bumpers (approx). | - 18' 0" | (548.64 cms) |
| Overall width, over wings (approx). | - 6' 5" | (195. 5 cms) front and rear |
| Overall height, unladen (approx). | - 6' 0" | (182. 8 cms) |
| Turning circle diameter (to outside edge of tyre) | - 45' 5" | (1,384.30 cms) R.H.and L.H. locks. |

NOTE: The overall length of the car, (both short and long wheelbase chassis) will vary according to the body and type of bumpers fitted. In cases where cars are intended for export, bumpers may vary considerably, thus affecting the overall length of the cars.

Chassis Weights: (Short Wheelbase Chassis). Complete with standard tyres 7.50" x 16.00", all accessories, tools, etc. and standard transmission.

COACHWORK: (Silver Wraith, Cont'd).

Dry Weight. R.H. drive, with Wilmot
Breedon heavy type bumpers.                  - 2,925 lbs. (1,326.8 Kgs).
Dry Weight. L.H. drive, with Wilmot
Breedon heavy type bumpers.                  - 2,955 lbs. (1,340.4 Kgs).

Chassis Weight. (Short Wheelbase)
Chassis). Complete with standard
tyres 7.50" x 16.00", all accessories,
tools, etc. and automatic transmission.

Dry Weight. L.H. drive, with Wilmot
Breedon heavy type bumpers.                  - 3,035 lbs. (1,376.6 Kgs).

Chassis Weights. (Long Wheelbase
Chassis). Complete with standard
tyres 7.50" x 16.00", all accessories,
tools, etc. and standard transmission.

Dry Weight. R.H. drive, with Wilmot
Breedon heavy type bumpers.                  - 3,000 lbs. (1,360.8 Kgs).
Dry Weight. L.H. drive, with Wilmot
Breedon heavy type bumpers.                  - 3,030 lbs. (1,374.4 Kgs).

Chassis Weight. (Long Wheelbase
Chassis). Complete with standard
tyres 7.50" x 16.00", all accessories,
tools, etc. and automatic transmission.

Dry Weight. L.H. drive, with Wilmot
Breedon heavy type bumpers.                  -. 3,110 lbs. (1,410.7 Kgs).

NOTE:  The above chassis weights are
       liable to slight variation due
       to individual chassis fittings.

COACHWORK: SILVER DAWN

DIMENSIONS. (With Standard Saloon all
Steel Body). Series A, B, C & D.

Wheelbase                                    - 10' 0"  (304.80 cms).
Track, front                                 - 4' 8¾"  (144.14 cms).
Track, rear                                  - 4' 10⅝" (148.90 cms).
Overall length, including bumpers
  (Wilmot-Breedon type).                     - 16' 0"  (487.68 cms).
Overall length, including bumpers
  (Pyrene type).                             - 16' 4½" (499.11 cms).
Overall width (over wings):-
  Front                                      - 5' 9"   (175.26 cms).
  Rear                                       - 5'10"   (177. 8 cms).
  Rear, with rear wing cover panels          - 5'11"   (180.34 cms).
Overall height:-
  Unladen                                    - 5' 6"   (167.64 cms).
  Laden                                      - 5' 4½"  (163.83 cms).
Turning circle dia. (To outside
             edge of tyre).                  - 41' 2"  (1,254.76 cms) R.H. and L.H.
                                                                      locks.

# SERVICE HANDBOOK

### SILVER WRAITH — SILVER DAWN — BENTLEY MK. VI.
### R. TYPE BENTLEY — PHANTOM IV.

COACHWORK: (Silver Dawn, Cont'd).

WEIGHT: (With Standard Saloon all
Steel Body). Series A, B, C & D.

| | | |
|---|---|---|
| Kerbside weight, R.H. drive | - 4,235 lbs. (1,921 Kgs) | These weights |
| Kerbside weight, L.H. drive | - 4,295 lbs. (1,948 Kgs) | are liable to |
| Dry weight, R.H. drive | - 4,040 lbs. (1,832 Kgs) | slight |
| Dry weight, L.H. drive | - 4,095 lbs. (1,857 Kgs) | variation to |
| | | individual |
| | | chassis |
| | | fittings. |

- The term "Kerbside" weight means
  that the car is complete with all
  accessories, petrol (18 galls.
  Imperial), oil and water, but less
  passengers.
  "Dry" weight means that the car is
  complete with all accessories, but
  less petrol, oil and water.

DIMENSIONS: (With Standard Saloon all
Steel Body). Series E and onwards.

| | |
|---|---|
| Overall length, including bumpers (Wilmot Breedon type) | - 16' 7½" (506.73 cms). |
| Overall length, including bumpers (Pyrene type) | - 16' 11½" (516.89 cms). |
| Overall length, with heavy export type bumpers | - 17' 6" (533.4 cms) approx. |

WEIGHT: (With Standard Saloon all
Steel Body and Standard Transmission.)

| | | |
|---|---|---|
| Kerbside weight R.H. drive | - 4,220 lbs. | (1,914 Kgs). |
| Kerbside weight L.H. drive | - 4,230 lbs. | (1,918 Kgs). |
| Dry weight R.H. drive | - 4,025 lbs. | (1,826 Kgs). |
| Dry weight L.H. drive | - 4,050 lbs. | (1,837 Kgs). |

WEIGHT: (With Standard Saloon all
Steel Body and Automatic Transmission).

| | | |
|---|---|---|
| Kerbside weight, L.H. drive | - 4,325 lbs. | (1,962 Kgs). |
| Dry weight L.H. drive | - 4,130 lbs. | (1,873 Kgs). |

CHASSIS WEIGHTS: (Complete with
Standard Tyres, all Accessories,
Tools, etc. and Standard Transmission).

| | | |
|---|---|---|
| Dry Weight. R.H. drive, with Wilmot Breedon export bumpers. | - 2,745 lbs. | (1,245.1 Kgs). |
| Dry Weight. L.H. drive, with Wilmot Breedon export bumpers. | - 2,770 lbs. | (1,256.5 Kgs). |

NOTE:    When Pyrene Bumpers are
         fitted this increases the
         weight by 5 lbs. (2.27 Kgs).

### SILVER WRAITH — SILVER DAWN — BENTLEY MK. VI.
### R. TYPE BENTLEY — PHANTOM IV.

CHASSIS WEIGHTS. (Complete with
Standard Tyres, all Accessories,
Tools, etc. and Automatic Transmission).

Dry Weight. L.H. drive, with Wilmot
Breedon export bumpers.                    - 2,850 lbs. (1,292.7 Kgs).
When Pyrene bumpers are fitted this
increases the weight by 5 lbs.
(2.27 Kgs).

NOTE:-   The above chassis weights are
         liable to slight variation due
         to individual chassis fittings.

COACHWORK:

   BENTLEY

DIMENSIONS. (With Standard Saloon all
Steel Body). Chassis Series A to P
inclusive.

| | | |
|---|---|---|
| Wheelbase | - 10' 0" | (304.80 cms) |
| Track, front | - 4' 8½" | (144.14 cms) |
| Track, rear | - 4'10⅞" | (148.90 cms) |
| Overall length, including bumpers (Wilmot-Breedon type) | - 16' 0" | (487.68 cms) |
| Overall length, including bumpers (Pyrene type) | - 16' 4½" | (499.11 cms) |
| Overall width (over wings):- | | |
| Front | - 5' 9" | (175.26 cms) |
| Rear | - 5'10" | (177.8 cms) |
| Rear, with rear wing cover panels | - 5'11" | (180.34 cms) |
| Overall height:- | | |
| Unladen | - 5' 6" | (167.64 cms) |
| Laden | - 5' 4½" | (163.83 cms) |
| Turning circle dia. (To outside edge of tyre). | - 41' 2" | (1,254.76 cms) R.H. and L.H. locks. |

WEIGHT: (With Standard Saloon all
Steel Body). Series A to P
inclusive.

| | | | |
|---|---|---|---|
| Kerbside weight, R.H. drive | - 4,210 lbs. | (1,909 Kgs) | These weights |
| Kerbside weight, L.H. drive | - 4,250 lbs. | (1,928 Kgs) | are liable to |
| Dry Weight  R.H. drive | - 4,015 lbs. | (1,821 Kgs) | slight |
| Dry Weight  L.H. drive | - 4,055 lbs. | (1,839 Kgs) | variation due |
| | | | to individual |
| | | | chassis |
| | | | fittings. |

The term "Kerbside" weight means that
the car is complete with all accessories,
petrol (18 galls. Imperial), oil and
water, but less passengers.
"Dry" weight means that the car is
complete with all accessories but
less petrol, oil and water.

COACHWORK: (Bentley, Cont'd).

DIMENSIONS: (With Standard Saloon all Steel Body). R Series and onwards).

| | | |
|---|---|---|
| Overall length, including bumpers (Wilmot Breedon type) | - 16' 7½" | (506.73 cms) |
| Overall length, including bumpers (Pyrene type) | - 16' 11½" | (516.89 cms) |
| Overall length, with heavy export type bumpers. | - 17' 6" | (533.4 cms) approx. |

WEIGHT: (With Standard Saloon all Steel Body and Standard Transmission).

| | | | |
|---|---|---|---|
| Kerbside weight, R.H. drive | - 4,215 lbs. | (1,912 Kgs) | These weights |
| Kerbside weight, L.H. drive | - 4,270 lbs. | (1,937 Kgs) | are liable to |
| Dry Weight R.H. drive | - 4,015 lbs. | (1,821 Kgs) | slight |
| Dry Weight L.H. drive | - 4,070 lbs. | (1,846 Kgs) | variation due to individual chassis fittings. |

WEIGHT: (With Standard Saloon all Steel Body and Automatic Transmission).

| | | |
|---|---|---|
| Kerbside weight, L.H. drive | - 4,365 lbs. | (1,980 Kgs) |
| Dry Weight L.H. drive | - 4,150 lbs. | (1,882.5 Kgs) |

CHASSIS WEIGHTS: (Complete with Standard Tyres, all Accessories, Tools, etc. and Standard Transmission.)

| | | |
|---|---|---|
| Dry Weight. R.H. Drive, with Standard type bumpers. | - 2,740 lbs. | (1,242.8 Kgs) |
| Dry Weight. L.H. Drive, with Wilmot Breedon export bumpers. | - 2,790 lbs. | (1,265.5 Kgs) |
| Dry Weight. R.H. Drive, with Pyrene Bumpers. | - 2,765 lbs. | (1,254.2 Kgs) |

NOTE:- When Pyrene bumpers are fitted to L.H. Models, this increases the weight by 5 lbs. (2.27 Kgs.)

CHASSIS WEIGHTS: (Complete with Standard Tyres, all Accessories, Tools, etc. and Automatic Transmission).

| | | |
|---|---|---|
| Dry Weight. L.H. drive, with Wilmot Breedon export bumpers. | - 2,870 lbs. | (1,301.8 Kgs). |
| Dry Weight. L.H. drive, with Pyrene bumpers. | - 2,875 lbs. | (1,304.1 Kgs). |

NOTE:- The above chassis weights are liable to slight variation due to individual chassis fittings.

# Service Handbook

COACHWORK:

### BENTLEY CONTINENTAL SPORTS SALOON.

DIMENSIONS:

| | | |
|---|---|---|
| Overall length, including bumpers | - 17' 2½" | (524.51 cms) |
| Overall length, with heavy export type bumpers. | - 17' 7½" | (537.2 cms) approx. |
| Overall width (over wings). | - 5'11½" | (181.61 cms) |
| Overall height (unladen). | - 5' 3" | (160.02 cms) |

WEIGHT: (Complete car).

| | | | |
|---|---|---|---|
| Dry weight. R.H. drive | - 3,610 lbs. | (1,637 Kgs) | These weights |
| Dry Weight. L.H. drive | - 3,635 lbs. | (1,649 Kgs) | are liable to |
| Kerbside weight. R.H. drive | - 3,805 lbs. | (1,726 Kgs) | slight |
| Kerbside weight. L.H. drive | - 3,830 lbs. | (1,737 Kgs) | variation due to individual chassis fittings. |

CHASSIS WEIGHTS: (Complete with all Accessories, Tools, etc. and Standard Transmission, less Radio and Heater.)

| | | |
|---|---|---|
| Dry Weight. R.H. drive with Standard Wilmot-Breedon bumpers. | - 2,740 lbs. | (1,242.8 Kgs) |
| Dry Weight. L.H. drive with Standard Wilmot-Breedon bumpers. | - 2,765 lbs. | (1,254.2 Kgs) |

NOTE:- The above chassis weights are liable to slight variation due to individual chassis fittings.

COACHWORK:

### PHANTOM IV.

DIMENSIONS:

| | | |
|---|---|---|
| Wheelbase | - 145" | (368.3 cms) |
| Track, front | - 58.4" | (148.3 cms) |
| Track, rear | - 63.2" | (160.5 cms) |
| Turning circle (to outside edge of tyre). | - 49' 1" | (1,496 cms) R.H. and L.H. locks. |

## PERIODIC LUBRICATION AND ADJUSTMENT.

### SCHEDULES "A" "B" & "C".

The three Schedules "A", "B" and "C" cover the whole vehicle and operate in the following manner :-

SCHEDULE "A" :-  To be carried out at the conclusion of every 5,000 miles, covers all the items associated with engine, chassis and coachwork requiring lubrication, cleaning and adjustment.

SCHEDULE "B" :-  To be carried out at the conclusion of every 10,000 miles. In addition to the repetition of the whole of Schedule "A", it covers the complete change of lubricant of all the main components, together with the inspection and rectification of those items not included at the lower mileage.

SCHEDULE "C" :-  To be carried out at the conclusion of every 20,000 miles. This schedule especially covers the requirements of periodic servicing of the Automatic Gearbox and only applies to cars fitted with Automatic Gearboxes.

### SCHEDULE "A".
### EVERY 5,000 MILES.

LUBRICATION:

1. Bonnet fasteners and locks.
2. Door locks and hinges and boot lid lock.
3. Ignition distributor shaft, contact breaker pivots and cam.
4. Steering column controls, accelerator, carburetter and clutch pedal mechanism.
5. Brake system pivot pins and bearings.
6. Propellor shaft universal joints (3 points) and sliding joint (1 point).

OIL LEVEL CHECKS:

1. Steering box.
2. Chassis lubrication tank.
3. Carburetter automatic air valve guides.
4. Lockheed master cylinder.
5. Gearbox.
6. Rear Axle.
7. Front and rear shock absorbers.
8. Drain and refill crankcase. Renew oil filter element.

### SILVER WRAITH — SILVER DAWN — BENTLEY MK. VI.
### R. TYPE BENTLEY — PHANTOM IV.

ENGINE & CHASSIS ADJUSTMENTS:

1. Check coolant level and top up if required ( When climatic conditions warrant, check specific gravity of coolant and advise owner if additi... anti-freeze is required).
2. Check fan belt tension. Adjust if necessary.
3. Check and reset inlet tappet clearances.
4. Clean sparking plugs. Check and reset gaps.
5. Clean contact breaker points. Reset gaps check and reset ignition ti...
6. Clean carburetter air valves (Bentley)
7. Check functioning of fuel pump (disconnect electrical leads and check pump independently).
8. Check and adjust clutch pedal free movement.
9. Adjust brakes and servo.
10. Check for excessive leakage at any point in the central chassis lubri... system.
11. Check and adjust tyre pressures.
12. Clean oil bath air filter element if fitted and refill with oil.

ELECTRICAL SYSTEM:

1. Check battery acid level. Top up with distilled water if required. Clean re-vaseline and tighten battery terminals.
2. Check complete electrical system for correct functioning.

ROAD TEST:

1. Test car on road.

### SCHEDULE "B".
### EVERY 10,000 MILES.

1. Repeat Schedule "A".
2. Drain and refill gearbox (Synchro-mesh type only)
3. Drain and refill rear axle.
4. Check starter motor reduction gear oil level and refill if required.
5. Clean carburetter air filter element.
6. Clean the two fuel strainers.

    1. The main fuel filter on the chassis cross member just forward of the petrol tank.

    2. The filter gauze in the carburetter float chamber feed connection.

7. Remove, check, and refit petrol pumps.

### SCHEDULE "C".
### EVERY 20,000 MILES.
### AUTOMATIC GEARBOXES ONLY.

1. Repeat Schedule "B".
2. Drain and refill gearbox (Automatic type). Check oil pressure. Clean oil breather in top of dipstick.
3. Check band adjustment. Test car on road.

# SECTION
# B
# ENGINE

# SERVICE HANDBOOK

SILVER WRAITH — SILVER DAWN — BENTLEY MK. VI.

R. TYPE BENTLEY — PHANTOM IV.

SECTION B.

# ENGINE

List of Illustrations:-

### SILVER WRAITH — SILVER DAWN — BENTLEY MK. VI.
### R. TYPE BENTLEY — PHANTOM IV.

## ENGINE.

| | |  |  |
|---|---|---|---|
| SILVER WRAITH | : | A to E Series | F Series Onwards |
| SILVER DAWN | : | A & B Series | C Series Onwards |
| BENTLEY | : | A to L Series | M Series Onwards |

| | | | |
|---|---|---|---|
| Number of cylinders | - | 6. In line. | 6. In line. |
| Bore | - | 3⅜". | 3⅝". |
| Stroke | - | 4½". | 4½". |
| Piston displacement | - | 4,256 c.c's. | 4.566 c.c's. |
| Compression ratio | - | +6.4:1 | +6.4:1 |
| H.P. | - | 29.4. | 31.54. |
| Sump capacity | - | 2 gallons. | 2 gallons. |
| Sparking plugs | - | 14 m/m. | 14 m/m. |
| Firing order | - | 1,4,2,6,3,5. | 1,4,2,6,3,5. |
| Weight | - | 640 lbs. approx. | |
| Suspension | - | 2 point. | 2 point. |

+ See cylinder head types.

The larger bore engine incorporates "Full-Flow" oil filtration instead of "By-pass" filtration, and is easily recognised by the large oil filter, and also the hinged-type oil filler cap on the rocker cover.

PHANTOM IV :

| | | |
|---|---|---|
| Number of cylinders | - | 8. In line. |
| Bore | - | 3⅜". |
| Stroke | - | 4½". |
| Piston displacement | - | 5,675 c.c's. |
| Compression ratio | - | 6.4:1. |
| H.P. | - | 39.2. |
| Sump capacity | - | 2¼ gallons. |
| Sparking plugs | - | 14 m/m. |
| Firing order | - | 1,6,2,5,8,3,7,4. |
| Weight | - | 790 lbs. approx. |
| Suspension | - | 2 point. |

The crankcase and cylinders are a cast-iron mono-bloc casting, the top portion of the cylinders being fitted with short "Bricrome" inserts, except early series which were "Flash Chromed".

The detachable cylinder head is of aluminium alloy.

Overhead inlet, and side exhaust valves are fitted, the inlet valve seats being of nickel chrome steel.

SILVER WRAITH — SILVER DAWN — BENTLEY MK. VI.

R. TYPE BENTLEY — PHANTOM IV.

## CYLINDER BLOCK.

| | | |
|---|---|---|
| Material | - | Cast Iron |
| Standard cylinder bore dia. | - | $3.5005"^{+1}$ and $3.625"^{+1\frac{1}{2}}$ |
| Cylinder bore tolerance (requiring re-bore) | - | .005" ovality |

## RECONDITIONING BORES:

The best method of determining the condition of the cylinders in an engine before reconditioning is the use of a dial gauge.

To use, the dial gauge is inserted in the cylinder bore, see Fig. B1. It is then turned spirally or completely rotated at as many points as may be desired, taking readings at each point.

If the dial gauge is set to correspond with the exact diameter of the standard bore, it is easy to determine the oversize piston as well as the amount of metal which must be removed from the cylinder walls.

Any reliable type of boring bar may be used. The operating instructions issued by the manufacturer should be followed explicitly.

The cylinder bores should be re-bored in stages of plus .005" to suit oversize piston range.

After the cylinder bores have been re-bored to within .003" of the size desired, they should be finished or polished with an expanding type hone.

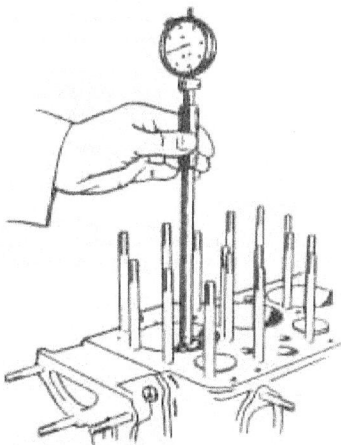

FIG. B1. CHECKING CYLINDER BORE.

| | | |
|---|---|---|
| Honing stones | | |
| (Rough) | - | Carburundum C-180-N-VGN |
| (Fine) | - | Carburundum C-320-N-VGN |
| Spindle Speed | - | 180 R.P.M. |
| Lubricant | - | Duckham's Aquicut H.M. Soluble oil, with paraffin |
| Honing | - | 33⅓% oil to 66 2/3% paraffin |
| Boring | - | 70 - 75% paraffin |

In operating, the hone is placed in the cylinder bore and expanded until it can just be turned by hand. The hone is then operated, up and down, in the bore until it begins to run free. Then the expanding nut, on the top of the hone, is tightened and the hone is again passed up and down, in the bore, until it runs free.

It should be noted that the standard bore size is 3.5005" or 3.625" and that any oversize must be to the full size, e.g. if it is necessary to re-bore to fit .010" oversize pistons, the cylinder bore size must be 3.5105" or 3.635", it is not permissible to be on the minus side.

## BRICROME INSERTS:

Originally, the top 2.250" of the cylinder bores were "flash chromed" to a thickness of .0015" on the bore diameter. This was later superseded by the fitting of short "Bricrome" inserts.

Where reconditioning is necessary to the original "chromed" bores, it will be necessary to bore out and fit the short inserts.

These inserts are available with a wall thickness of .062" and are intended to cover re-boring up-to and including the maximum standard oversize of plus .040".

If the lower part of a cylinder is badly scored, i.e., too deeply to clean up to match with the remaining bores, a three-quarter length liner is available for fitting to the lower portion.

To fit inserts, remove cylinder head studs and set up the crankcase absolutely true with the original cylinder bore.

Check the new inserts for parallelism at the ends and correct if necessary. (See Fig.B2).

The bottom face must be machined perfectly true. Any slight discrepancy on this end will cause a gap on one part of the "butt" joint.

When boring out for fitting the .062" inserts, adjust the boring cutters so that the first cut is heavy enough to get underneath the original chromium plating. Counterbore to a depth relative to the length of the new insert.

Measure the outside diameter of the inserts in six places and taking the mean diameter of these measurements, bore out the cylinder block, allowing for .0025" to .0035" interference between the insert and the cylinder bore.

Prior to pressing in the inserts, check chamfer at bottom of counterbores is smaller than that on inserts, to ensure clearance and no interference on the butt joints when inserts are fully pressed home.

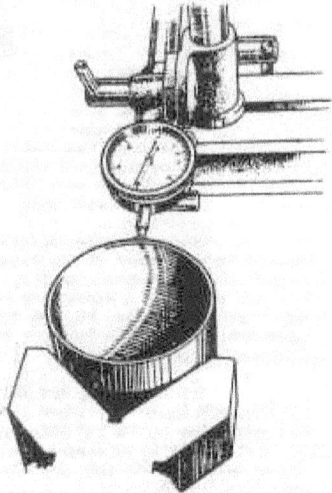

FIG. B2.  CHECKING END FACE OF INSERT.

Press in the inserts DRY, with hydraulic press, pressure required approximately 3 tons.

If fitting two inserts to the one cylinder bore, press in lower one and upper one together from the top.

Grind off ends of inserts flush with top surface of crankcase.

Bore straight through inserts, or inserts and original cylinder bores to the nearest standard oversize, allowing .003" for honing.  Replace the cylinder head studs.

Hone bores to required size, taking care that, owing to the differences in texture of the insert and bore, to guard against scoring.

CORE PLUGS:

Steel core plugs with an aluminium joint washer are now standard for the crankcase and in a case where one is defective and leaking, it should be replaced with one of .062" oversize on thread diameter.

Care must be taken when removing the defective plug, not to damage the female threads in the block;  after removal, clean up threads with suitable tap, and screw in new plug, using a little "Heldite", with a two pin spanner.

### CONNECTING RODS.

| | | |
|---|---|---|
| Material | - | Chrome Molybdenum Steel. |
| Length | - | 7.950" centre to centre. |
| Journal size | - | 1.999" (Standard). |
| Bearing size | - | Pre-sized bore to suit diameter of C/pin. |
| Running clearance | - | .0015" - .003". |
| End play (rod-piston) | - | .004" - .014". |
| Small end bush (fixed) | - | .75025" + $\frac{1}{4}$ bore (reamed in position). |
| Weight (with cap) | - | 1 lb. 12 ozs. approx. |

The connecting rod should be checked for alignment of the piston and gudgeon pin with big-end bearing, for both bend and twist, on a connecting rod alignment fixture, see Fig. B3, and corrected as necessary.   Allowable twist or bend in alignment is .001".

The connecting rod bolts are of the "fitted" type;  if, after cleaning up the bolt holes in the rod and cap, the bolts are found to be slack, the holes should be reamed out and .004" oversize bolts are fitted.

The connecting rod bolt nuts should be pulled down using a box spanner and a six inch tommy bar.

The connecting rod should be installed on the crankshaft so that the oil spray hole is facing towards the "thrust"

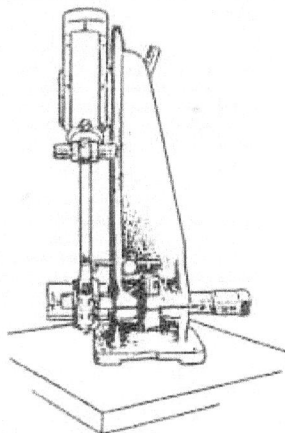

FIG. B3. CHECKING CON-ROD ALIGNMENT.

side of the engine.

On production, both the rod and the cap are etched with the number of the cylinder to which they are to be assembled. The numbers are etched on the "non-thrust" side. When the rods are being re-assembled they should be placed back into the same cylinder from which they were removed with the etched numbers opposite the "thrust" side.

The connecting rod big end bearings are of the pre-sized steel backed strip type, having a copper and lead-indium bearing facing. They should be discarded and new ones fitted if the lead-indium plating has worn through on any part.

## PISTONS.

| | | |
|---|---|---|
| Piston material | - | Aluminium alloy. |
| " type | - | Split skirt. |
| " length | - | 4.270". |
| Clearance top | - | .020". |
| " bottom | - | No clearance on thrust dia. |
| Groove depth | | |
| Compression | - | .129". |
| Oil scraper | - | .1425". |
| Number of Rings | | |
| Compression | - | 2. |
| Oil scraper | - | 1. |
| Ring Width | | |
| Compression | - | .176". |
| Oil scraper | - | .092". |
| Ring gap (all). | - | .014". |
| Gudgeon pin dia. | - | .750". |
| " " length | - | 3.475". |
| " " fit in piston | - | .0002" interference, cold. |
| " " fit in rod | - | Running fit. |

## FITTING PISTONS:

The pistons are cam ground to the desired eccentricity at the top of the skirt and this eccentricity is progressively reduced so that it is substantially less at the bottom of the skirt. This gives a more uniform bearing with minimum friction.

Oversize pistons are normally available in steps of .005" up to .040" oversize. The gudgeon pins are individually fitted to each piston and should be kept strictly in combination.

In fitting pistons it is most important that the correct clearance, .003" - .0035", at the top of the skirt is obtained, and the only reliable method is to measure the cylinder bore with a dial gauge and the piston with a suitable micrometer.

Fitting should be done at normal room temperature, 70°F.

## PISTON RINGS:

To check the ring gap, insert the piston ring into the cylinder bore about 1" from the top and measure gap with feeler gauge. See Fig.B4.

If the ring gaps are less than the recommended clearance, remove the ring and with a very fine file, dress the ends until the correct clearance is obtained.

Side clearance of ring in groove is measured with a feeler gauge with the ring fitted in groove but not installed on piston. The back side of the ring should be rolled entirely around the groove, to ensure that the ring is free and does not bind at any point.

The oil scraper and intermediate compression rings are marked "TOP" on the upper side of the ring and they must be installed with the marked side towards the top of the piston.

The rings should be installed in sequence over the top of the piston. Do not remove rings over the piston skirt, as this is liable to cause scoring.

The gaps in the three rings should not be in a vertical line, neither should there be any ring gap over the gudgeon pin. It is therefore desireable to stagger the gaps so that they are equally spaced around the circumference of the piston.

FIG. B4. CHECKING RING GAP.

ASSEMBLING:

The piston should be assembled to the rod so that the slotted side of the piston is opposite to the oil spray hole in the connecting rod and also the thrust side of the cylinder.

The gudgeon pins having a slight interference fit, assembly will be eased if the piston is first immersed in boiling water for a short period.

In slipping the piston back into the cylinders, use extreme care and do not force the rings into the bore. Compress the rings with a Ring Clamp until they enter the cylinder freely.

## CRANKSHAFT AND MAIN BEARINGS.

### CRANKSHAFT.

|  |  | SILVER WRAITH | : |
|  |  | SILVER DAWN | : |
|  |  | BENTLEY | : |

| Material | - | Nitrided chrome molybdenum steel |
|---|---|---|
| Number of journals | | 7 |
| Journal size | - | 2.750" - $\frac{1}{2}$ (standard diameter) |
| Journal length | - | No. 1 - 1.825" |
| | | Nos.2,3,5 and 6 - 1.400" |
| | | No. 4 - 1.875" |
| | | No. 7 - 2.100" |
| Number of crankpins | - | 6 |
| Crankpin size | - | 1.999" - $\frac{1}{2}$ standard diameter |
| Balance weights | - | Detachable |
| Crankpin length | - | 1.375" |
| Thrust washer (Overall thickness) | - | .092" - 2 |

### SILVER WRAITH — SILVER DAWN — BENTLEY MK. VI.

### R. TYPE BENTLEY — PHANTOM IV.

NOTE:- With the introduction of the 3.625" bore engine the crankshaft end web
were strengthened, which means that the crankshaft for the 3.500" bore
engine is not interchangeable with the shaft for the 3.625" bore engine.
Also, the Front and Rear main bearing shells have also been reduced in
width to suit, thus rendering these also non-interchangeable.

3.625" bore engine.

| | | |
|---|---|---|
| Journal length | — Front | 1.450" |
| | — Rear | 1.850" |
| Width of Main Bearing Shells | — Front | 1.118" |
| | — Rear | 1.468" |

PHANTOM IV :

| | |
|---|---|
| Material | — Nitrided chrome molybdenum steel |
| Number of Journals | — 9 |
| Journal size | — 2.750" - ½ (standard diameter) *Bearing housing 2.9175 (Tunnel)* |
| Journal length | — No. 1 - 1.9375" |
| | Nos.2,3,4,6,7 and 8 - 1.400" |
| | No. 5 - 1.875" |
| | No. 9 - 2.100" |
| Number of crankpins | — 8 |
| Crankpin size | — 1.999" - ½ *Bearing housing 2.1415 inch* |
| Crankpin length | — 1.375" |
| Balance weights | — Integral with webs |
| Thrust washer (Overall thickness) | — .092" - 2 |

Alternative crankshafts may be used on production, either "EN.19" or
"Hykro" steel. This specification is always stamped on the front web or flange
and must be checked for the "Hardness" figure of the nitriding.

EN.19 - 570 V.P.N. min.
Hykro - 800 V.P.N. min.

The crankshaft is
balanced statically twice, once
as a detail crankshaft only,
and again with oil caps and
flywheel fitted. It is
supported in steel backed
copper-lead-indium split type
shell bearings, which are
detachable and can be renewed
in certain cases without the
necessity of removing the
crankshaft.

End thrust is taken
by the two alloy metal split
thrust washers, arranged either
side of the centre main bearing.
The lower halves of these washers
are keyed to the bearing cap to
prevent rotation. Note:Front end
washers are etched "X" on
original assembly.

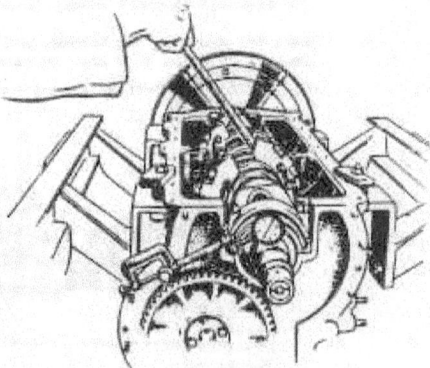

FIG. B5.   CHECKING CRANKSHAFT END FLOAT.

SILVER WRAITH — SILVER DAWN — BENTLEY MK. VI.

R. TYPE BENTLEY — PHANTOM IV.


The crankshaft end float is:- .002" - .006". (Measured on centre journal.)

INSPECTING THE CRANKSHAFT:

Before inspecting the crankshaft for wear or bowing, check whether the journals or crankpins have previously been re-ground.

1.  Mount the crankshaft on a suitable stand and check with a michrometer for wear on the journals and crankpins.
    Any wear of the journals must be taken into account when checking for bow.

2.  Mount the six-throw shaft with the journals, Nos. 1 & 7 in a pair of vee-blocks placed on a marking out table, and ensure that the crankshaft axis is parallel with the table and free to rotate.

    Before mounting crankshaft, it would be advisable to use a test bar on the vee-blocks to ensure that the crankshaft will be parallel with the table. If a test bar is not available, the journals, Nos. 1 & 7 should be checked for diameter, and after the crankshaft is mounted on the vee-blocks, a second check should be made with a dial height gauge over the journals. If the axis is not parallel, packing pieces should be inserted under the vee-blocks.

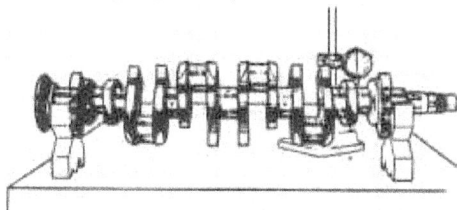

FIG. B6. CHECKING PARALLELISM OF SHAFT AND TABLE.

The eight-throw shaft should be mounted on journals 5 and 9.

3.  Turn the shaft in the blocks, and test for bowing on the centre journal by means of a dial indicator mounted on a scribing block.

    The maximum permissible bow is:- .010".

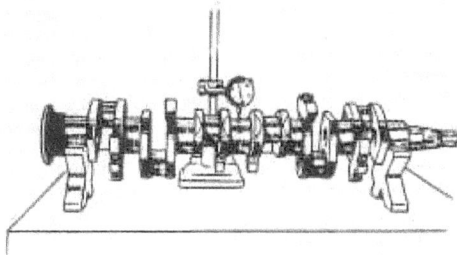

FIG. B7. CHECKING CRANKSHAFT BOW.

The errors due to ovality of the journals must be taken into account in arriving at the figure for bowing which will be half the maximum dial reading of the indicator.

4.    Next, turn the shaft so that the webs of each crankpin are at first parallel, and then at right angles to the table and check the parallelism of the crankpins in each instance.

In checking the crankpins and journals for wear, the fact that the shaft may have been reground, should be taken into account.

If it is found that it is necessary to regrind the journals or crankpins, a hardness test must be made to establish whether the shaft should be renitrided before returning to service. This test should be made on a crankpin or journal that has had the greatest amount removed from its diameter by the grinding operation, or, in the case of a seizure, on that part of the bearing most affected by the heating (this can usually be seen by the "temper" colours remaining on the bearing surface.)

The minimum hardness figure for reground crankshafts is either 570 or 800 V.P.N., see page B7, using a 10 kg. load for the Vickers Diamond Pyramid Machine; and a crankshaft that has been subjected to bearing seizure should be rejected if less than this figure is registered.

REGRINDING AND LAPPING:

As previously stated, the journals and crankpins of a crankshaft which have become scored or rendered outside the limits of ovality, may be restored to a satisfactory working condition by regrinding, lapping and polishing. The combined operations should result in the diameter of the journals or crankpins being reduced in multiples of .010" to a maximum of .060" undersize.

Grinding should not be resorted to, unless it is necessary to remove more than .005" to restore a true diameter. Up to this amount, the journal or pin should be rectified by lapping only.

When regrinding journals, no metal should be removed from the crank-web. For details of the stages of regrinding, see table below:-

| STAGES OF REGRINDING. | DIAMETER OF JOURNALS. | DIAMETER OF CRANK-PINS. |
|---|---|---|
| 1 | 2.740 - $\frac{1}{4}$ | 1.989 - $\frac{1}{4}$ |
| 2 | 2.730 - $\frac{3}{4}$ | 1.979 - $\frac{3}{4}$ |
| 3 | 2.720 - $\frac{3}{4}$ | 1.969 - $\frac{3}{4}$ |
| 4 | 2.710 - $\frac{1}{2}$ | 1.959 - $\frac{1}{2}$ |

The shaft should be set up on the grinding machine, using an adaptor for the flanged end, and a centring plug for the other end when grinding the journals.

SILVER WRAITH — SILVER DAWN — BENTLEY MK. VI.

R. TYPE BENTLEY — PHANTOM IV.

Proceed to run the machine, centring the end journals to as fine a limit as possible, and checking by means of a suitably mounted dial indicator. Grind the journals 2 or 7 true, and adjust the supporting blocks of the machine into position under either journal. Proceed to grind the shaft, working from the end according to which journal has been made true. The grinding wheel must not be allowed to touch the journal before the crankshaft is thoroughly wetted with the grinding lubricant, and this should be fed in liberally on the ingoing side of the wheel. In order to avoid cracking, it is desirable that arrangements should be made to heat the lubricant and maintain it at a temperature of 65 degrees to 69 degrees C. On no account must the grinding wheel touch the side radii of the crank-web. Stops must be arranged on the machine to limit the travel of the grinding wheel to approximately 0.010" on each face. The radius of the wheel should be carefully controlled to ensure that the grinding fades out not more than half-way round the radius.

Grinding should be to .001" of finished size.

When regrinding the crankpins, the crankshaft is set up on the throw-blocks which form the normal equipment of the grinding machine. The same precautions as for the journals should be observed, and the centre pair of crankpins ground first. Having reground the necessary pins and/or journals, a final operation should be effected whereby the front face of the driving flange is ground true. This should only necessitate a light skimming with the grinding wheel.

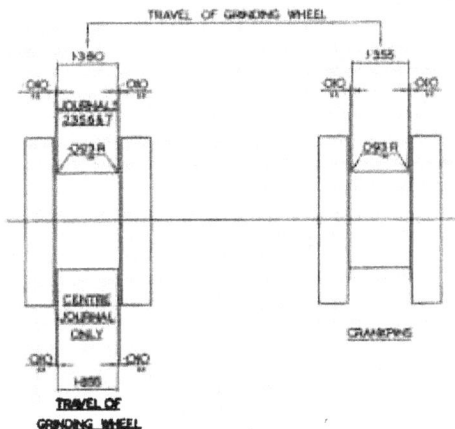

FIG. B8. GRINDING WHEEL.

Various coolants or lubricants may be employed in regrinding the crankshaft. Of these, water with the addition of soda in the proportion of 1 lb. to 50 gallons of water and sufficient soap solution to produce a mixture which can be frothed easily is the simplest. A recommended grinding wheel for the above operation is a 80K. Upon completion of the grinding operation, the crankshaft should be tested magnetically for cracks with the greatest care, ensuring that all electrical connections are carefully made, and the crankshaft supported on proper clamps to obtain a good electrical connection to the magnetizing apparatus.

In no case should upper and lower half bearings be interchanged, and care must be taken to see that the locating lips correctly register in their recesses.

On new engines, or engines to be fitted with reground crankshafts and undersize main bearings and correct size thrust washers, the correct clearances are:-

        Diametrical clearance - .002" - .0035".
        End float           - .002" - .006".

## TO RENEW A BEARING WITHOUT REMOVING CRANKSHAFT:

1.  Remove the lower half.
    It is not necessary to remove the oil pump assembly, although it is easier if this is done. It must be remembered that if the oil pump is removed the ignition will need retiming.

2.  Unlock the nuts retaining the main bearing caps, slacken off, but do not remove.

3.  Remove the cap of the bearing to be renewed.
    If more than one bearing is to be renewed only one cap is to be removed at a time.

4.  Slide the top half of the bearing out, around the crankshaft journal using a thin strip of flexible steel. The bearing should be pushed out in the direction of rotation of the crankshaft.

    The new half bearings may be inserted in the opposite direction.

    The thrust washers may be renewed in the same way.

    Note the size stamped on the back of the removed bearing and replace accordingly.

5.  Remove the lower half of the bearing from the bearing cap and replace the corresponding half of the new bearing.

    NOTE:- It is essential that all parts concerned should be thoroughly clean and oiled with clean engine oil before assembly.

## MAIN BEARING INSPECTION:

(i)   The main bearings and thrust washers should be removed from the crankcase and their respective caps, and thoroughly washed in paraffin.

(ii)  Visually inspect. Reject bearings obviously damaged, or those showing any wearing through of the lead indium plating.

      A range of pre-finished undersize bearings are available in steps of .005" for use with reground crankshafts.

      No reaming is necessary, if the crankshaft has been reground, say .015", undersize from standard, a set of .015" undersize bearings will give the correct running clearance.

      To check the bearing fit, all bearings must be fitted and bolted down, when the crankshaft should be free enough to turn by hand.

In no case should upper and lower half bearings be interchanged, and care must be taken to see that the locating lips correctly register in their recesses.

On new engines, or engines to be fitted with reground crankshafts and undersize main bearings and correct size thrust washers, the correct clearances are:-

Diametrical clearance - .002" - .0035".
End float - .002" - .006".

## TO RENEW A BEARING WITHOUT REMOVING CRANKSHAFT:

1. Remove the lower half.
   It is not necessary to remove the oil pump assembly, although it is easier if this is done. It must be remembered that if the oil pump is removed the ignition will need retiming.

2. Unlock the nuts retaining the main bearing caps, slacken off, but do not remove.

3. Remove the cap of the bearing to be renewed.
   If more than one bearing is to be renewed only one cap is to be removed at a time.

4. Slide the top half of the bearing out, around the crankshaft journal using a thin strip of flexible steel. The bearing should be pushed out in the direction of rotation of the crankshaft.

   The new half bearings may be inserted in the opposite direction.

   The thrust washers may be renewed in the same way.

   Note the size stamped on the back of the removed bearing and replace accordingly.

5. Remove the lower half of the bearing from the bearing cap and replace the corresponding half of the new bearing.

   NOTE:- It is essential that all parts concerned should be thoroughly clean and oiled with clean engine oil before assembly.

## MAIN BEARING INSPECTION:

(i) The main bearings and thrust washers should be removed from the crankcase and their respective caps, and thoroughly washed in paraffin.

(ii) Visually inspect. Reject bearings obviously damaged, or those showing any wearing through of the lead indium plating.

   A range of pre-finished undersize bearings are available in steps of .005" for use with reground crankshafts.

   No reaming is necessary, if the crankshaft has been reground, say .015", undersize from standard, a set of .015" undersize bearings will give the correct running clearance.

   To check the bearing fit, all bearings must be fitted and bolted down, when the crankshaft should be free enough to turn by hand.

# SERVICE HANDBOOK

### SILVER WRAITH — SILVER DAWN — BENTLEY MK. VI.
### R. TYPE BENTLEY — PHANTOM IV.

No filing or shimming of bearings or caps is permissible.

## CRANKCASE OIL FLOW CHECK:

When the crankshaft has been installed in an overhauled engine, and before proceeding with further rebuilding, a check should be made to ensure that there is a satisfactory oil flow to the main bearings, connecting rods and camshaft bearings.

All external outlets should be blanked off and a suitable oil pump to give approximately 30 lbs./sq.in. pressure, should be connected to the crankcase oil pump flange. Operate pump; slowly turn crankshaft and check oil flow to main bearings, connecting rod big-ends, connecting rod small ends and camshaft.

Remove and check size of spray hole in oil jet to camshaft centre gear, this should be .0748" + .003". If necessary, open out to suit using a No.48 morse drill.

### CRANKSHAFT DAMPER.

SILVER WRAITH :
SILVER DAWN :
BENTLEY :

Slipping Load - 14 lbs. + 1 lb. at 17½" radius.
(Drive assembled less driving springs)

Two types of Low Inertia Spring Drives are fitted, known as the "Wraith" and "Bench" types. The "Wraith" type must be fitted to the crankshaft in a partly assembled condition, whereas the "Bench" type can be assembled and offered up as a complete unit.

The device is a composite component and deals with two different and distinct conditions.

1. The Spring Drive - provides a flexible coupling between the crankshaft and the crankshaft pinion, maintaining a constant pressure at the driving teeth, irrespective of the variations in load, by means of radially disposed coil springs interposed between dog members which are attached to the crankshaft and pinion.

2. The Vibration Damper - deals with the torsional vibrations of the crankshaft by means of friction imposed on the inertia of a small flywheel, using cotton duck washers loaded by coil springs.

NOTE:- As the starter dog is attached to the floating part of the device, it is essential that the engine is turned from the rear at the main flywheel, in its normal direction of rotation, when timing the engine.

SILVER WRAITH — SILVER DAWN — BENTLEY MK. VI.

R. TYPE BENTLEY — PHANTOM IV.

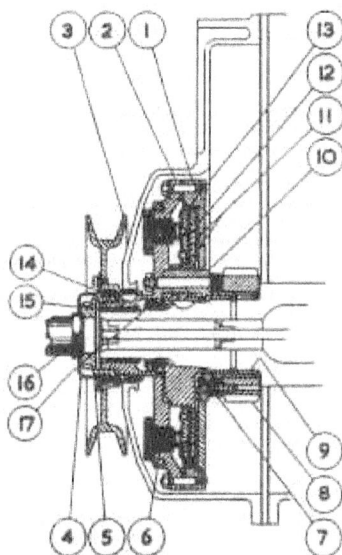

FIG. B9. "WRAITH" TYPE.

1. Damper Wheel (Rear)
2. Damper Wheel (Front)
3. Fan Pulley.
4. Star Locking Plate.
5. Withdrawal Nut (Pulley).
6. Damper Spring.
7. Distance Piece.
8. Crankshaft Pinion.
9. Bush (Rear).
10. Driven Dog.
11. Friction Washers.
12. Friction Drive.
13. Pressure Plate.
14. Serrated Nut.
15. Bush (Front).
16. Starter Dog.
17. Serrated Nut.

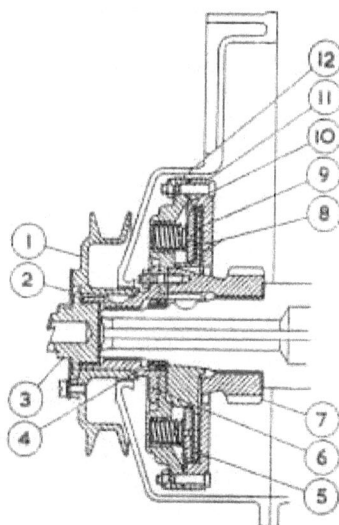

FIG. B10. "BENCH" TYPE.

1. Fan Pulley.
2. Locking Plate.
3. Starter Dog.
4. Serrated Nut.
5. Damper Spring.
6. Hub.
7. Crankshaft Pinion.
8. Friction Washers.
9. Friction Drive.
10. Pressure Plate.
11. Damper Wheel (Rear).
12. Damper Wheel (Front).

## DERANGEMENT OF DAMPER:

Mechanical failures are unlikely, inefficiency is mainly due to lengthy periods of idleness, such as car storage, resulting in the cotton duck washers adhering to the friction plates, producing engine vibration periods noticeably at 2,500 engine r.p.m., half the torsional period of the crankshaft.

## COTTON DUCK WASHERS:

It is essential that new cotton duck washers are fitted at each overhaul period as these become ingrained with carbon and superficially glazed.

### SILVER WRAITH — SILVER DAWN — BENTLEY MK. VI.
### R. TYPE BENTLEY — PHANTOM IV.

The new washers should be immersed in oil, (S.A.E. 20), overnight and afterwards placed under a press for approximately twelve hours to flatten.

They should then be assembled and "ironed" by movement of the damper wheels backwards and forwards to reduce to a uniform thickness, especially at the scarf joint. Afterwards check by using spring balance as illustrated.

FRICTION FACES:

Check all friction faces for scoring and regrind or turn as necessary.

The sectional drawing, (Fig.B12) illustrates the limits to which they may be safely ground or turned.

PHANTOM IV:

The rubber tuned harmonic balancer consists of a small flywheel, attached to and driven by the crankshaft through a friction disc, mounted over five rubber bushed bolts.

With the engine running, any torsional vibrations of the crankshaft are damped out by the movement of the flywheel on the rubber mounted bolts.

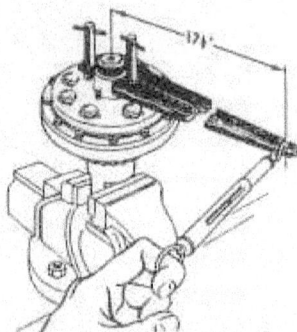

FIG. B11. CHECKING SLIPPING LOAD OF DAMPER.

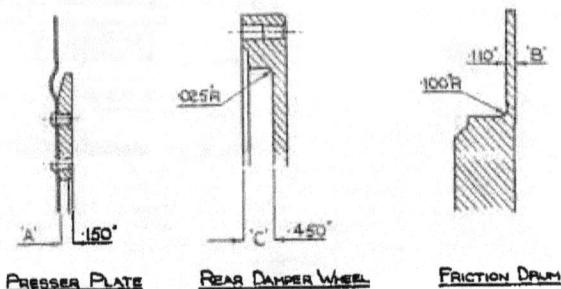

PRESSER PLATE        REAR DAMPER WHEEL        FRICTION DRUM

FIG. B12.   GRINDING LIMITS.

NOTE:- The starter dog is attached directly to the end of the crankshaft and therefore in this case, turning of the engine for timing purposes may be done by means of the starting handle.

1. Pulley.
2. Damper Springs.
3. Friction Disc.
4. Hub.
5. Starter Dog.
6. Flywheel Bush.
7. Rubber Bush.
8. Flywheel.

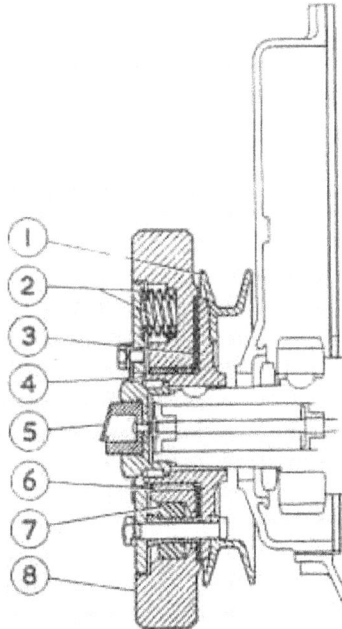

FIG. B13. HARMONIC BALANCER.

## CAMSHAFT.

SILVER WRAITH :
SILVER DAWN :
BENTLEY :

| | | |
|---|---|---|
| Drive | — | Helical toothed gear. |
| Bearings | — | Four Babbit lined steel shells. |
| End thrust | — | Spring loaded button to plate on Wheelcase. |
| End float | — | .002" - .006". |
| Bearing clearance | — | .002" - .003". |

Valve Timing — Silver Wraith and Silver Dawn.
    Inlet opens — 12° after T.D.C.
    Inlet closes — 35° after B.D.C.
    Exhaust opens — 44° before B.D.C.
    Exhaust closes — 6° before T.D.C.
Valve Timing — Bentley.
    Inlet opens — $3\frac{1}{2}$° after T.D.C.
    Inlet closes — $43\frac{1}{2}$° after B.D.C.
    Exhaust opens — $40\frac{1}{2}$° before B.D.C.
    Exhaust closes — $1\frac{1}{2}$° before T.D.C.
Camshaft journal
    diameter — 1.998" - $\frac{1}{2}$.

## CAMSHAFT.

Two types of cam formation have been employed on the Bentley camshaft, namely the "High-Lift" and "Low-Lift". As a means of identification the "High-Lift" Part No. is RE.5157, this has now been superseded and all models are fitted with the long duration "Low-Lift" type.

In the event of a new camshaft being required for an engine previously fitted with a "High-Lift" type, a "Low-Lift" type should be fitted, and in such case, new inlet valve springs, will be required.

Whenever a camshaft is removed from an engine, the journals should be checked with a micrometer for ovality, this must not exceed .001".

Next, mount the camshaft on "Vee" blocks and check the alignment by means of a dial indicator, if this is more than .002" the camshaft should be straightened.

NOTE:- During the straightening operation, care must be taken to protect the centre journal to prevent damage to its surface.

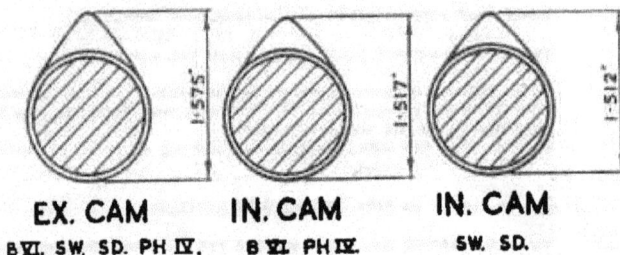

EX. CAM    IN. CAM    IN. CAM
B VI. SW. SD. PH IV.   B VI PH IV.   SW. SD.

FIG. B14. CAM MEASUREMENTS.

A further check should be made of the wear on the cams, see Fig. B14.

The wear limit is .016" if this is exceeded the camshaft must be replaced.

CAMSHAFT BEARINGS:

All the camshaft bearings are pressed into the crankcase and accurately line reamed at the time of assembly. The bearings are lubricated through holes which line up with the oil passages from the main bearings.

To remove, replace and line ream the camshaft bearings in service, a special set of tools is necessary. See Fig. B15.

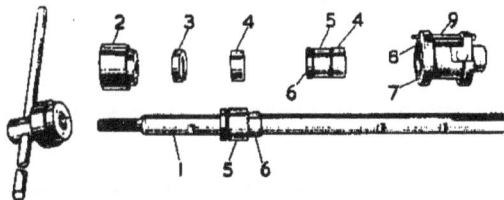

FIG. B15. CAMSHAFT BEARING TOOLS.

| | |
|---|---|
| 1. Arbor | 6. Bearing Support Piece |
| 2. Front Support | 7. Rear Support |
| 3. Locating Washer | 8. Locating Dowel |
| 4. Bearing Retainer | 9. Retaining Bolts |
| 5. Camshaft Bearing | |

METHOD OF OPERATION:

1. Remove camshaft bearing cover from rear facing.

2. Mount rear support piece in position, note dowelling.

3. Mount front support piece in position and insert bar.

4. Mount bearing support piece on bar complete with bearing and clip in required position. The bearing bar is slotted and dowelled in position on the support pieces.
The bearings are dowelled onto the bearing supports to ensure correct location of oil holes.

5. Rotate handle to draw bearing into position.

6. Withdraw bearing support piece and fit "Martell" cutter, ream through to correct size.

Camshaft bearing bore - 2.000" + $\frac{1}{2}$.

CAMSHAFT GEARS:

Two types of driving gears have been used, originally a "Fabric" gear was fitted and this has now been superseded by an "Aluminium" gear. In any case of failure of an original "Fabric" gear, replacement should be by the "Aluminium" gear.

SILVER WRAITH — SILVER DAWN — BENTLEY MK. VI.

R. TYPE BENTLEY — PHANTOM IV.

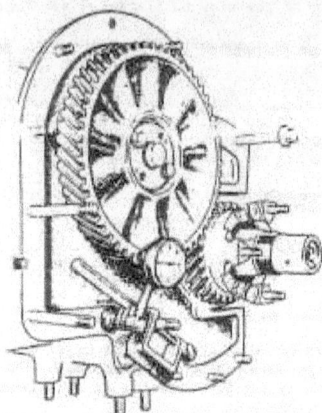

FIG. B16.  CHECKING RUN-OUT OF CAMSHAFT        FIG. B17.  CHECKING PINION BACKLASH.
GEAR.

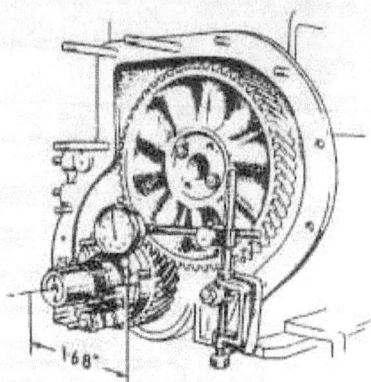

NOTE:- When the "Fabric" gear is replaced by the "Aluminium" gear it will
also be necessary to replace the crankshaft pinion as these are
paired on a rig to give the correct backlash .002" - .005".

ASSEMBLY:

Check that the three lock plates on front cover for camshaft location
plate are locked up before fitting cam wheel.  Also, check spring is assembled
in plunger before fitting.

PHANTOM IV :

| | | |
|---|---|---|
| Drive | - | Helical toothed gear |
| Bearings | - | Six Babbit lined steel shells |
| End thrust | - | Spring loaded button to plate on wheelcase |
| End float | - | .002" - .006" |
| Bearing clearance | - | .002" - .003" |
| Valve Timing | | |
| Inlet opens | - | $3\frac{1}{2}^{\circ}$ after T.D.C. |
| Inlet closes | - | $43\frac{1}{2}^{\circ}$ after B.D.C. |
| Exhaust opens | - | $40\frac{1}{2}^{\circ}$ before B.D.C. |
| Exhaust closes | - | $1\frac{1}{2}^{\circ}$ before T.D.C. |
| Camshaft journal dia.- | | $1.998" - \frac{1}{2}$. |

CAMSHAFT:

The camshaft is of the long duration low-lift type and the same remarks
apply with regard to checking as for the Silver Wraith and Bentley camshafts.

CAMSHAFT BEARINGS:

Except that there are six bearings instead of four, the previous remarks apply.

CAMSHAFT GEARS:

The driving gear is of the aluminium type.

### ENGINE TIMING.

### ALL MODELS.

The valve timing should be set with the engine cold.

1. Set and lock the inlet valve rocker on No.1 cylinder to .030".

2. Turn camshaft in normal direction of rotation until the inlet valve in No.1 cylinder just commences to open. Turn the crankshaft, see NOTE, under Crankshaft Damper, until the I.O. marking on the flywheel coincides with the timing pointer fixed to the housing.

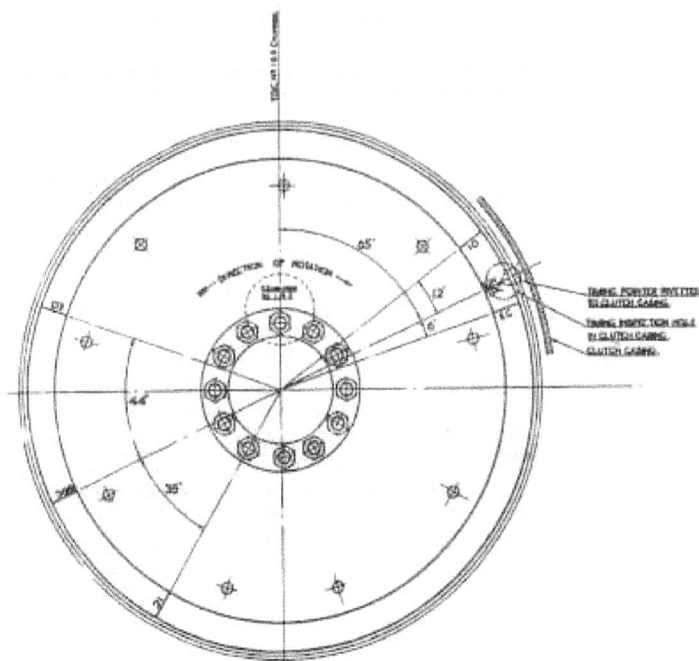

FIG. B18. VALVE TIMING.

On later series cars, a rationalised form of flywheel marking is incorporated. Flywheels are only marked with the T.D.C. position and four marks either side of this position designated 5, 10, 15, 20° Early or Late.

The valve timing remains the same, the Bentley and Phantom IV should be set on T.D.C. and the Silver Wraith and Silver Dawn on the 10° Late marking.

NOTE:- Reference to Fig.B18, IGN/TDC marking on flywheel is 65° before or after True TDC of Nos. 1 and 6, owing to positioning of inspection hole on clutch casing to suit Right or Left-hand chassis.

3.   Adjust the camshaft gear on the hub to mate with the crankshaft pinion.

The hub of the crankshaft gear is provided with either four or eight holes. One tooth on the camshaft gear is equivalent to a movement of 1.75" approx: measured on the flywheel periphery, and it should be possible to obtain a final valve timing to within $\frac{7}{8}$", i.e., $\frac{1}{2}$ tooth setting, or 7/16", i.e. $\frac{1}{4}$ tooth setting, of the flywheel marking, according to whether four or eight hole camshaft gear is fitted.

NOTE:- In actual practice, owing to the crankshaft pinion and spring drive unit being keyed on to the shaft, and the camshaft gear being retained by studs to the camshaft, the above operation will be somewhat difficult procedure. Therefore, it is recommended that the studs in the camshaft should be temporarily withdrawn and replaced with setscrews. Care must be taken to replace the studs when the correct position has been found.

It is preferable that the timing should be on the "late" side rather than "early".

4.   On completion, reset and lock the inlet valve rocker to .006".


### WHEEL CASE.

Crankshaft bore dia. - 2.4375" + 2.
Acme thread dia.    - 2.425" - 2.
Clearance           - .006" - .008".

The wheelcase is dowelled to the crankcase by three fitting bolts, except in the case of certain early models where two taper dowels are used.

It is essential that the correct clearance, as above, is maintained, and this should be checked with a feeler gauge, using an alignment sleeve.

# SERVICE HANDBOOK

## VALVES AND TAPPETS.

| | |
|---|---|
| Seat Angle | - 45° |
| Valve guides | - Inlet - cast iron. |
| | - Exhaust - bronze. |
| Valve clearance in guide | |
|     Inlet | - .0015" - .003". |
|     Exhaust | - .0035" - .0055". |
| Spring pressure | |
|     Inlet inner, compressed to | - 1.300", 12 - 17 lbs. |
|     Inlet outer, compressed to | - 1.600", 40 - 48 lbs. |
|     Exhaust, compressed to | - 1.525", 40 - 48 lbs. |
| Valve rocker clearances | |
|     Inlet | - .006"⎫ |
|     Exhaust | - .012"⎬ Cold. |

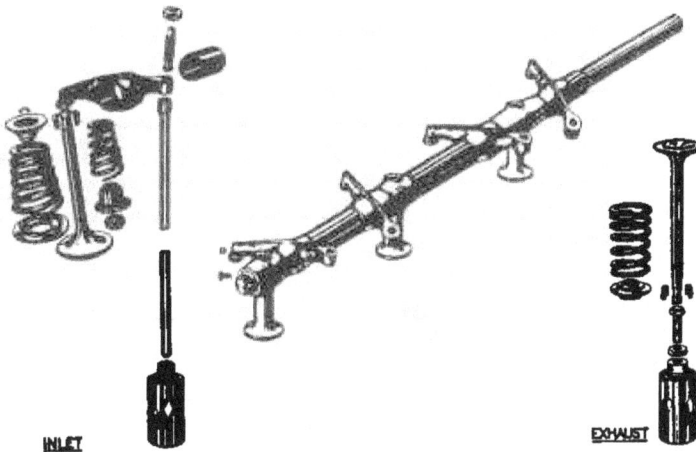

INLET              EXHAUST

FIG. B19.  VALVE ROCKER MECHANISM.

## VALVE GUIDES:

The clearance between valve guides and valve stems is most important. Lack of power and noisy valves may be due to worn guides.

The inlet valve guides should be checked with a new inlet valve and similarly the exhaust valve guides should be checked with a new exhaust valve owing to the difference in stem diameters.

Cast iron inlet valves guides are pressed into the cylinder head and these as supplied are .331" internal diameter, and must be reamed in position to .3437" + ¼.

The bronze exhaust valve guides require a special tool for drawing into the crankcase. These guides as supplied, are .370" + 2 internal diameter, and must be reamed in position to .3755" + ½.

Replacement guides are .002" oversize on the outside diameter to retain the correct interference fit.

## VALVES:

The exhaust valves have "Stellited" facings, a special heat resisting metal.

When valves are removed, the valve stems and heads should be cleaned on a buffing wheel to remove all carbon, etc., the condition of the valves and valve stems can then be checked.

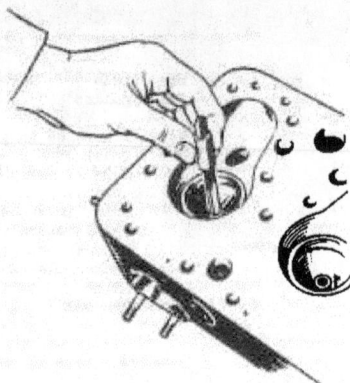

FIG.B20.   CHECKING INLET VALVE GUIDE
WITH PLUG GAUGE.

Valves that are pitted or burned, can be refaced on a standard valve grinding machine.  After refacing, machine to maintain dimension "A" and "B" (Figs. 21 and 22).   To check for perfect contact, apply a little Prussian Blue to seat and replace valve, give the valve a half turn to the right and then to the left, if the Blue is equally coated, the grinding is perfect.   On the other hand if the "Blueing" is uneven, indicating an uneven spot, the valve must be re-ground until it seats properly.

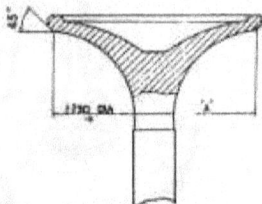

FIG. B21.   INLET VALVES.

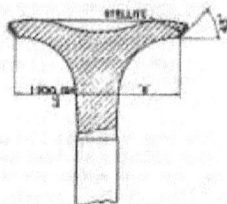

FIG. B22.   EXHAUST VALVES.

Check the valve stem diameters, with a micrometer, for wear.

| | | |
|---|---|---|
| Inlet valve stem diameter | .3422" | - .3416" min. |
| Exhaust valve stem diameter | .372" | - .37125" min. |

## VALVE TAPPETS:

The valve tappets are directly operated by the camshaft and are located slightly off-centre with the cams, which results in spinning the tappets so that the cams do not engage the same point each time the valves open.

The inlet and exhaust valve tappets are graded in steps of .00025" to give a selective fitting.   They should be a light push fit in the bore (dry).

The working clearance is .0005" - .001".

The exhaust valve clearance adjustment is on the bottom tappet.

VALVE SEATS:

      Inlet valve seat diameter    - 1.750" + 5.
      Exhaust valve seat diameter  - 1.625" + .15.

Reconditioning the valve seats is an important operation to ensure engine economy and operation. Extreme care should be used to maintain the stated limits and clearances.

The valve seat must ┝ round to the proper width and angle, good judgement must be used when nar. _ng a valve seat to ensure that the seat contacts the centre of the valve facing. This should be checked by smearing a little Prussian Blue on the valve facing and pressing the valve into its seating, without rotating it. A complete circle of marking should appear on both valve seat and facing.

On completion of these operations, great care must be taken to make certain that all grinding dust and foreign matter is removed from the valve, seatings, guides and ports of both the cylinder head and the cylinder block, and also the tappet chamber.

ROCKER ARMS AND SHAFTS:

The inlet rocker arms are handed being off-set to the right and left-hand. The left-handed being fitted to the odd numbered cylinders and the right-handed to the even numbered cylinders.

The rocker shafts are hollow, and have holes drilled in them to allow oil to pass into the rocker arm bushes. The rockers being drilled through to allow a small amount of oil to lubricate the push rod end and the contact point with the valve stems.

On the Phantom IV the rocker shaft is in two sections, one end of each rocker shaft is plugged, and they

FIG. B23.  CHECKING VALVE FACE AND SEATING.

must be installed on the cylinder head with the open ends to the centre pedestal.

The dome headed adjusting screws are screwed into the rocker arms and locked with locknuts. All adjustments to the inlet valve clearance is done at this point, no adjustment being provided at the bottom tappets.

CHANGING INLET VALVE SPRING IN SITU:

Remove rocker cover and rocker shaft assembly.

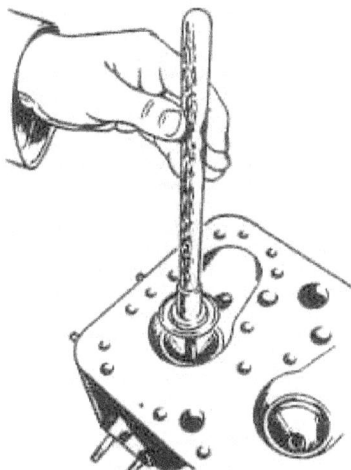

SILVER WRAITH — SILVER DAWN — BENTLEY MK. VI.

R. TYPE BENTLEY — PHANTOM IV.

1. Rocker Shaft.

2. Rocker Arm.

3. Ball-headed adjusting screw.

4. Lock nut.

5. Feeler gauge in position.

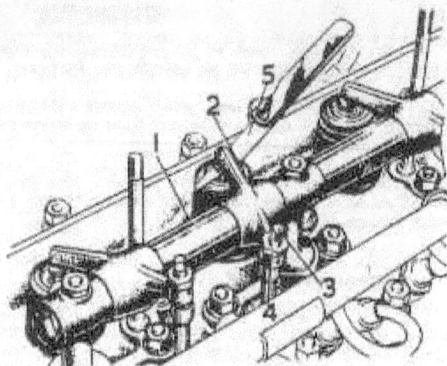

FIG. B24. ADJUSTING INLET VALVE ROCKER CLEARANCE.

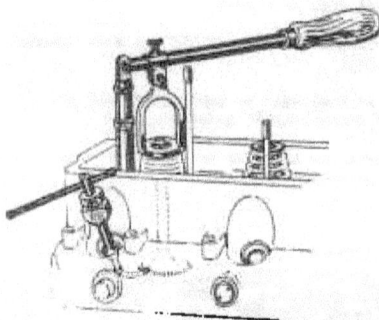

FIG. B25. INLET VALVE SPRING COMPRESSING TOOL.

Place the valve spring compressing tool and the valve holder in position as shown in Fig. B25. The locking nut in the valve holder should first be unscrewed to release the split taper collet which grips the spindle and the bent portion of the latter inserted through the sparking plug hole. Then, screw the holder into position.

The spindle should then be turned and simultaneously pulled away from the head, so that it is bearing in the hollow of the appropriate valve head of the spring to be removed. While holding in position, tighten locking nut.

Operate the valve spring compressing tool from a convenient stud and remove the valve wedges, release and withdraw top washer and springs. If valve stem packing is to be replaced, the small retaining ring will have to be removed from groove in valve stem.

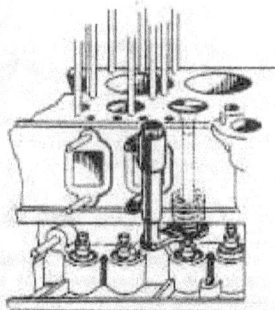

FIG. B26. EXHAUST VALVE SPRING COMPRESSING TOOL.

## CYLINDER HEAD.

The cylinder head is of aluminium, and great care must be taken when removing from the engine not to damage the facings.

If the cylinder head joint proves difficult to "break", this may be done by motoring over the engine with the starter motor, with the sparking plugs in position but not connected.

The return of high grade fuels to the home market has made possible the introduction of a new cylinder head (RE.19451) on the 3.625" bore engine only, giving a higher compression ratio.

This head when fitted with the copper and asbestos gasket (RE.14764) gives a compression ratio of 6.75:1.

It is possible to further increase the compression ratio to 7.12:1 by fitting a steel gasket (RE.16320) in place of the C/A gasket, but this modification must only be carried out after consultation with the London Service Department.

At present the Bentley Continental Sports Saloon is fitted with cylinder head (RE.16876) and a C/A gasket (RE.14764) giving a ratio of 7.27:1.   In no circumstances must the steel gasket be fitted with this head.

Future production will be with cylinder head (RE.19451) and steel gasket (RE.16320) giving a compression ratio of 7.20:1.

Whenever the head is removed, it is advisable to check and clean the oil way, in the head and cylinder block, to the centre rocker pedestal.

When decarbonising or reconditioning the cylinder head, only a blunt rounded tool or wire brush, should be used to remove the carbon.

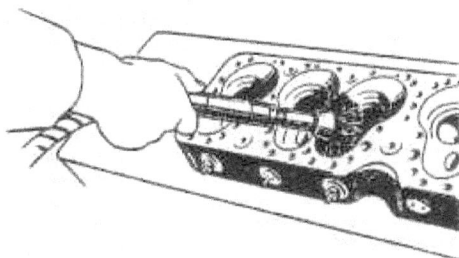

FIG. B27.   DECARBONISING CYLINDER HEAD.

When replacing the cylinder head, the nuts should be tightened in the order indicated.   They should be tightened down gradually, from the centre, a few turns at a time, so that the head is not distorted.

After the engine has been run for a while on test, the cylinder head nuts should be given a final tightening.

**SILVER WRAITH — SILVER DAWN — BENTLEY MK. VI.**

**R. TYPE BENTLEY — PHANTOM IV.**

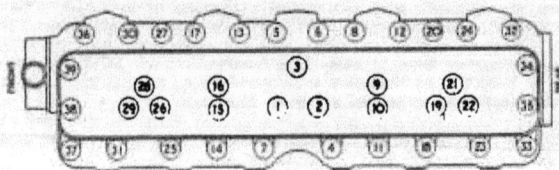

FIG. B28. ORDER OF TIGHTENING CYLINDER HEAD NUTS.
(6 cylinder engines).

CYLINDER HEAD GASKET:

Whenever the cylinder head is removed, the cylinder head gasket should be discarded and a new one fitted when the head is replaced.

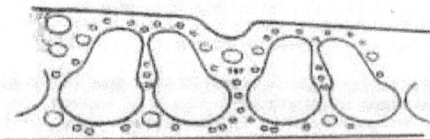

FIG. B29. CYLINDER HEAD GASKET.

The gasket is marked "TOP" on one side and this should be placed uppermost on the cylinder block to make contact with the cylinder head.

When replacing the C/A gasket, it is recommended that a little jointing compound is smeared along the exhaust side edge to ensure that no leak should occur on this side, the coolant connections being on the outside of the cylinder studs.

SPARKING PLUG ADAPTORS:

If the thread of the adaptor is defective, the adaptor may be removed and replaced with either a standard or oversize one as necessary.

The sparking plug adaptor has a .002" interference fit in the cylinder head, and this must be retained.

To remove the adaptor, unscrew the locking ring - left-hand thread - and, using a square tapered drift, a few light taps should ensure a good "bite", unscrew the insert with a suitable spanner. The latter having a right-hand thread.

To replace adaptor, clean up female thread in cylinder head, then place head in oven at a temperature of 300°F for one hour. Coat the male thread of the insert with Whitmore's Compound No.5, or similar grease. Remove the head from the oven and screw the insert into position, right-hand thread. Face off if necessary. Afterwards, screw in the locking ring, using a suitable peg spanner.

When the cylinder head has cooled, clean up threads and measure with suitable micrometer, select insert to give the required interference fit.

Place cylinder head in oven at a temperature of 300°F for one hour. Coat threads of insert with Whitmore's Compound No.5, or similar type of grease. Remove cylinder head from oven and screw in insert.

Machine off the spannering head on valve seat insert, and lock seat in position by centerpopping.

The seat should now be ground and finished as described on page B.24.

## ENGINE LUBRICATION.

Capacity - 16 pints, Silver Wraith.
"    "      Silver Dawn.
"    "      Bentley.
18   "      Phantom IV.

## TYPE:

The engine lubrication is supplied by a positively driven gear pump mounted in the lower half of the crankcase with an external oil relief valve unit to control the maximum pressures to the high-pressure and low-pressure systems.

The system provides positive high-pressure lubrication to all the main, connecting rod and camshaft bearings, and also a low-pressure supply to the overhead rocker shaft and the timing gear.

The high-pressure supply is approximately 25 lbs./sq.in.
The low-pressure supply is approximately  5 lbs./sq.in.

Two types of lubrication are employed, namely "By-pass" and "Full-flow" filtration.  See diagrams (Figs. B.30 and B.31).

On all chassis previous to:-

Silver Wraith - F Series.
Silver Dawn  - C Series.
Bentley      - M Series.
Phantom IV   - B Series.

the system used was "By-pass", i.e., a bleed from the high-pressure oil gallery was taken through the external by-pass oil filter.  (See Fig. B.30).

On the above series and onwards, full-flow filtration is employed, i.e., the oil is delivered direct from the pump to an external "full-flow" filter and from there returned to high-pressure system.  (See Fig. B.31).

## OIL PUMP:

The oil pump is a positive gear type and runs at half the engine R.P.M. It consists of two spur gears enclosed in a housing; in operation, oil is drawn from the crankcase sump through a floating fine mesh screen.  The oil then passes through a pipe to the oil pump and from there to the oil distribution system.

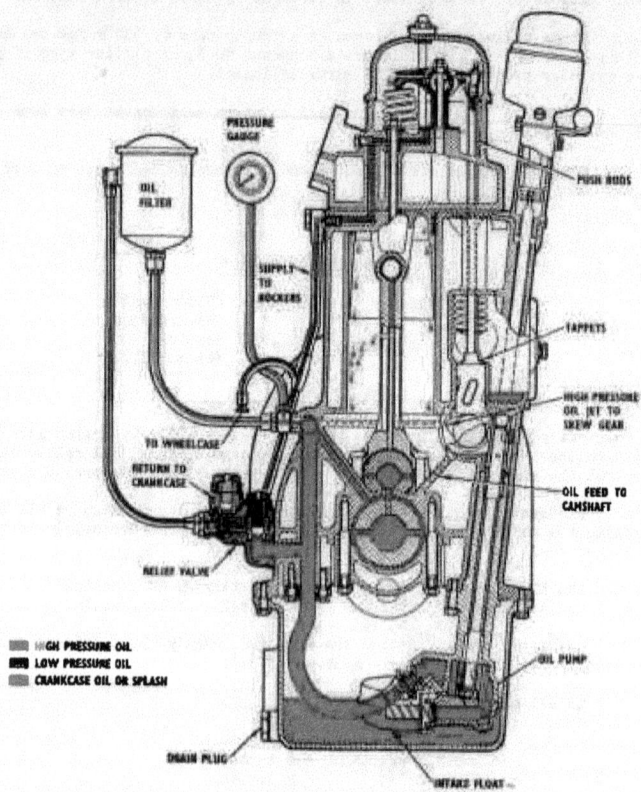

PRESSURE GAUGE

OIL FILTER

PUSH RODS

SUPPLY TO ROCKERS

TAPPETS

HIGH PRESSURE OIL JET TO SKEW GEAR

TO WHEELCASE

RETURN TO CRANKCASE

OIL FEED TO CAMSHAFT

RELIEF VALVE

HIGH PRESSURE OIL
LOW PRESSURE OIL
CRANKCASE OIL OR SPLASH

OIL PUMP

DRAIN PLUG

INTAKE FLOAT

FIG. B.30   ENGINE LUBRICATION SYSTEM (BY-PASS)

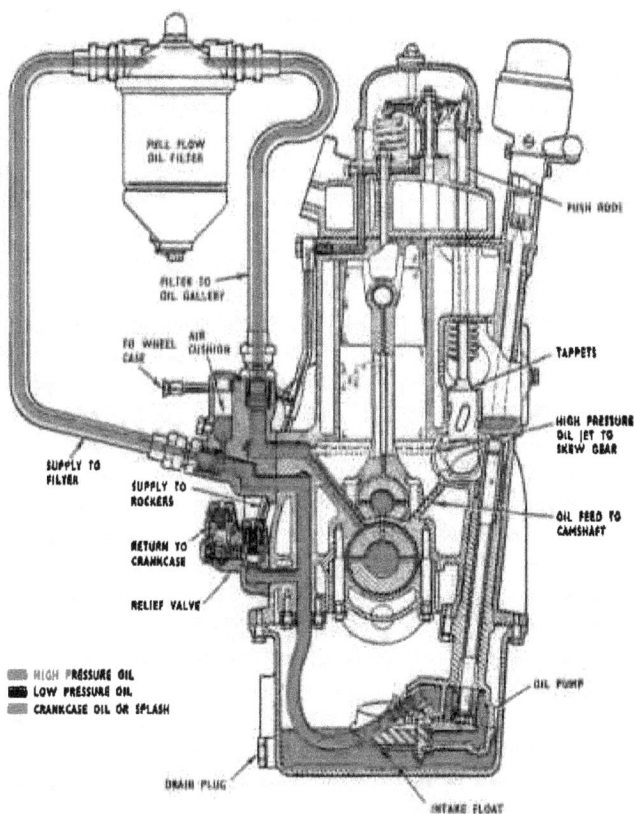

FULL FLOW
OIL FILTER

PUSH RODS

FILTER TO
OIL GALLERY

TO WHEEL    AIR
CASE      CUSHION

TAPPETS

HIGH PRESSURE
OIL JET TO
SKEW GEAR

SUPPLY TO
FILTER

SUPPLY TO
ROCKERS

OIL FEED TO
CAMSHAFT

RETURN TO
CRANKCASE

RELIEF VALVE

HIGH PRESSURE OIL
LOW PRESSURE OIL
CRANKCASE OIL OR SPLASH

OIL PUMP

DRAIN PLUG

INTAKE FLOAT

FIG. 8.31.—ENGINE LUBRICATION SYSTEM (FULL-FLOW)     R 662

SILVER WRAITH — SILVER DAWN — BENTLEY MK. VI.

R. TYPE BENTLEY — PHANTOM IV.

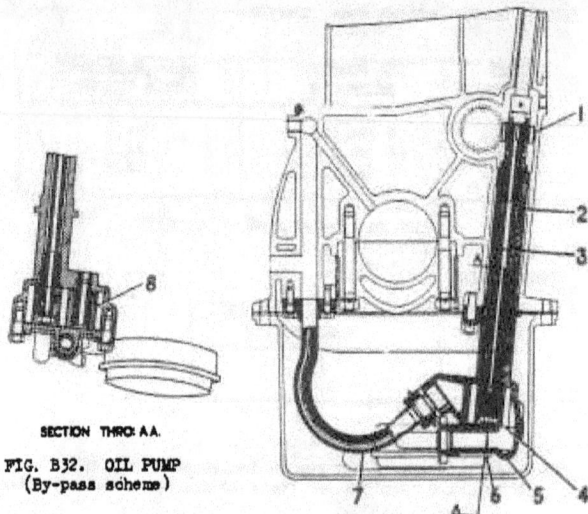

SECTION THRO: A.A.

FIG. B32. OIL PUMP
(By-pass scheme)

1. Driven gear, upper.
2. Oil pump casing.
3. Oil pump driving shaft.
4. Driving gear, lower.

5. Oil pump cover.
6. Pump float.
7. Oil delivery pipe.
8. Oil pump driven gear.

To dismantle the oil pump it will be necessary to remove the ignition distributor, as a combined drive from the camshaft operates both units.

1. Remove the distributor cover and set engine with rotor arm in firing position for No.1 cylinder, i.e. No.1 piston is at T.D.C. with both valves closed.

2. Remove the distributor and housing from the cylinder block. DO NOT slacken the distributor clamping plate screw as this would disturb the timing.

3. Note and mark the angular positions of the "flats" on the tongue of the distributor driving shaft for replacement on re-assembly.

4. Disconnect and withdraw the oil pump assembly complete with drive shaft.

SILVER WRAITH — SILVER DAWN — BENTLEY MK. VI.

R. TYPE BENTLEY — PHANTOM IV.

The oil pump should be rig flow-tested as under:-

SILVER WRAITH, SILVER DAWN, BENTLEY:-

| PUMP R.P.M. | OIL PRESSURE RESTRICTED | MIN. ACCEPTANCE FLOW PTS. MIN. |
|---|---|---|
| 500 | 8 lbs./sq.in. | 11.2 |
| 1000 | 18 lbs./sq.in. | 26.4 |
| 1500 | 28 lbs./sq.in. | 42.6 |

Inlet Oil Temperature to be 90°C.

PHANTOM IV:-

| PUMP R.P.M. | MIN. ACCEPTANCE FLOW PTS. MIN. |
|---|---|
| 2000 | 68.0 |

Dis-assemble the pump and remove the gears. The fit of the gears in the housing, the backlash, and the end float should be checked as under:-

Fit of gears in casing    - .0025" - .005".
Backlash of gears    - .005" - .008" at 1.278" pitch diameter.
End float of gears    - .0025" - .005".

Re-assemble by reversing the above instructions.

<u>OIL RELIEF VALVES:</u>

Both valves are in series and are not adjustable, no attempt must be made to alter the spring settings by interfering with the springs or varying the number of washers under the plugs. The H.P. valve seat is slotted to ensure a supply of oil to the low-pressure system under all conditions of running.

H.P. valve spring, free length 1.8125" compressed to 1"   = 4½ lbs.
L.P. valve spring, free length 1.790" compressed to .900" = 4 ozs.

To dismantle the relief valves, remove the plugs and lift out valves, clean and examine the seats.

If there are any signs of pitting on the valve, remove unit and lightly lap valve onto its seat, using a mixture of Turkeystone Powder and thin oil.

After lapping, thoroughly wash, re-assemble, and refit.

SILVER WRAITH — SILVER DAWN — BENTLEY MK. VI.

R. TYPE BENTLEY — PHANTOM IV.

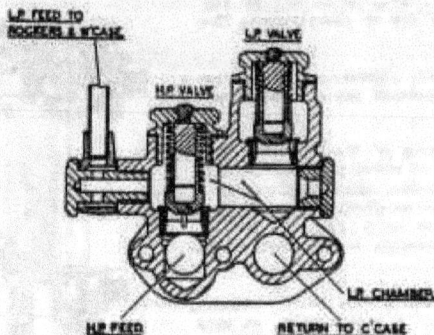

FIG. B33. OIL RELIEF VALVES.

BY-PASS OIL FILTER:

The oil filter is designed so that the element restricts the flow of the by-passed oil to ensure that the by-passing effect of the filter should no appreciably rob the main pressure system.

The element is non-demountable and therefore cannot be cleaned. The element should be discarded and a new one fitted every 10,000 miles.

FULL-FLOW OIL FILTER:

The oil filter is of the felt element type contained within a mesh canister. Oil enters from the top centre and passes through the element to the outside.

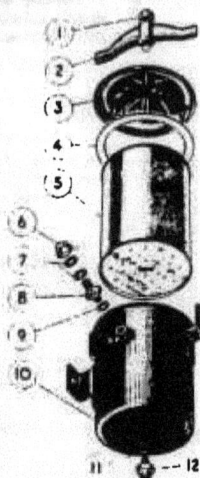

1. Yoke screw.
2. Yoke.
3. Cover.
4. Cover washer.
5. Element.
6.)
7.)
8.) Return connection.
9.)
10. Filter bowl.
11.)
12.) Feed connection.

FIG. B34. BY-PASS FILTER.

If the element becomes clogged with sediment, it will be lifted off its seating and so allow an unrestricted flow of oil.

The felt element should not be cleaned, but discarded and renewed every 5,000 miles.

Two types of "full-flow" filters are fitted, one on which the tank is held on by six set-screws, this model is now superseded by one on which the tank is held by one centre bolt fixing. The interior felt elements are interchangeable for both types.

To obtain access to the element, remove set-screws or centre bolt, as case may be, and lower tank containing filter canister. Remove canister and remove either the wing nut or the knurled nut, depending on type, from bottom face and remove bottom cover. Extract felt element and the two felt washers, top and bottom covers, fit new element and washers, refill bowl with oil and re-assemble.

1. Filter bowl.
2. Retaining screws.
3. Mesh canister, containing felt element.

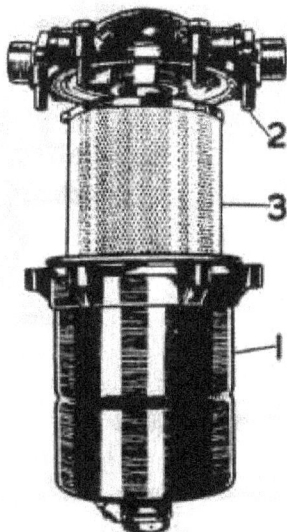

FIG. B35. FULL-FLOW OIL FILTER.

**SERVICE HANDBOOK**

SILVER WRAITH — SILVER DAWN — BENTLEY MK. VI.
R. TYPE BENTLEY — PHANTOM IV.

# SECTION
# C
# FUEL SYSTEM

# SERVICE HANDBOOK

### SILVER WRAITH — SILVER DAWN — BENTLEY MK. VI.
### R. TYPE BENTLEY — PHANTOM IV.

SECTION  C

FUEL SYSTEM

List of Illustrations:

## SILVER WRAITH — SILVER DAWN — BENTLEY MK. VI.
## R. TYPE BENTLEY — PHANTOM IV.

### FUEL SYSTEM.

Tank Capacity       - 18 or 23 gallons.
Fuel Delivery       - S.U. Twin Electric Pumps.
Air Cleaner type    - Mesh or Oil Bath.
Carburetter
   Silver Wraith
   Silver Dawn
   Bentley L.H.D.    - Stromberg, dual down-draught,
   Phantom IV       type AAV. 26M, or
   Bentley R.H.D.    Zenith, single down-draught type DBVC.42.
                 Two S.U. carburetters.
Fuel Pumps       - Delivery pressure, carburetter,
                 1 lb./sq.in. (Min. at tick-over.)

### STROMBERG JET DATA.

**SILVER WRAITH** - R.H. and L.H. drive.

| | A.C. Air Cleaner | | Oil Bath Air Cleaner | | |
| --- | --- | --- | --- | --- | --- |
| | Sea Level to 4,000 ft. | Over 4,000 ft. | Sea Level to 3,000 ft. | 3,000 ft. to 6,000 ft. | Over 6,000 ft. |
| Main Jets | .044" | 130 cc" | .041" | .040" | .039" |
| Power Jet | No. 54 | No. 54 | No. 57 | No. 57 | No. 57 |
| Idle Jet | No. 70 | No. 70 | No. 70 | No. 70 | No. 70 |
| Idle air bleed | No. 60 | No. 60 | No. 60 | No. 60 | No. 60 |
| Float needle seat | .101" | .101" | .101" | .101" | .101" |
| Accelerator pump capacity | 16-20 cc's | 16-20 cc's | 16-20 cc's | 16-20 cc's | 16-20 cc's |
| Vacuum kick-gap | .125" | .125" | .125" | .125" | .125" |

**SILVER DAWN** - R.H. and L.H. drive.
**BENTLEY**     - L.H. drive.

| | A.C. Air Cleaner | | Oil Bath Air Cleaner | | |
| --- | --- | --- | --- | --- | --- |
| | Sea Level to 4,000 ft. | Over 4,000 ft. | Sea Level to 3,000 ft. | 3,000 ft. to 6,000 ft. | Over 6,000 ft. |
| Main Jets | .045" | .041" | .042" | .041" | .040" |
| Power Jet | No. 54 | No. 54 | No. 57 | No. 57 | No. 57 |
| Idle Jet | No. 70 | No. 70 | No. 70 | No. 70 | No. 70 |
| Idle air bleed | No. 60 | No. 60 | No. 60 | No. 60 | No. 60 |
| Float needle seat | .101" | .101" | .101" | .101" | .101" |
| Accelerator pump capacity | 16-20 cc's | 16-20 cc's | 16-20 cc's | 16-20 cc's | 16-20 cc's |
| Vacuum kick-gap | .125" | .125" | .125" | .125" | .125" |

SILVER WRAITH — SILVER DAWN — BENTLEY MK. VI.

R. TYPE BENTLEY — PHANTOM IV.

PHANTOM IV -

A.C. Air Cleaner      Oil Bath Air Cleaner

|  | Sea Level to 4,000 ft. | Over 4,000 ft. | Sea Level to 3,000 ft. | 3,000 ft. to 6,000 ft. | Over 6,000 ft. |
|---|---|---|---|---|---|
| Main Jets | .046● |  |  |  |  |
| Power Jet | No. 54 |  |  |  |  |
| Idle Jet | No. 66 |  |  |  |  |
| Idle air bleed | No. 52 |  |  |  |  |
| Float needle seat | .101" |  |  |  |  |
| Accelerator pump capacity | 16-20 cc's |  |  |  |  |
| Vacuum kick-gap | .125" |  |  |  |  |

● denotes Broad Ring type.

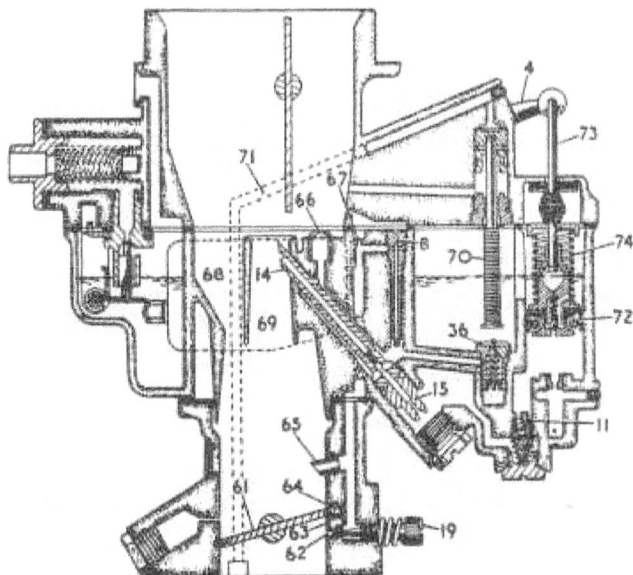

FIG. 01. DIAGRAMMATIC SECTION - STROMBERG CARBURETTER.

4. Lever accelerator pump.
8. Idle tube.
11. Pump check valve.
14. Main discharge tube.
15. Metering jet.
19. Idle valve screw.

36. Power by-pass jet.
61. Throttle valve.
62. Idle discharge hole.
63. First progression hole.
64. Second progression hole.
65. Secondary air bleed.

# SERVICE HANDBOOK

### SILVER WRAITH — SILVER DAWN — BENTLEY MK. VI.
### R. TYPE BENTLEY — PHANTOM IV.

66. High speed air bleed.
67. Primary air bleed.
68. Outer choke.
69. Inner choke.
70. Vacuum piston spring.

71. Vacuum channel.
72. Accelerator pump piston.
73. Accelerator pump rod.
74. Duration spring, accelerator pump.

## STROMBERG CARBURETTER ADJUSTMENT:

The carburetter is carefully adjusted to the engine before leaving the factory and the settings should not be altered unnecessarily.

There are two adjustments on the carburetter, one for the idling speed and one for the idling mixture. Both of these adjustments should be made together, and only when the engine is well warmed up.

FIG. C2.  CARBURETTER ADJUSTMENT.

1. Main jets.
2. Idling jets.
3. Throttle stop screw.
4. Choke control cam.
5. Fast idle cam.

(i) The throttle stop screw should be set so that the engine runs at approximately 350 r.p.m. The stop screw adjusts both throttle valves in unison, these being mounted on a common spindle.

(ii) Adjust each slow running mixture screw separately, to give the most even running. Each of these screws adjusts the mixture strength to the three, or four, cylinders supplied by each choke tube, turning the screw in weakens the mixture, and turning the screw out enrichening it. The approximate setting is 1¼ turns back from the closed position. Operate these screws only with the fingers.

## DIAGNOSIS OF FAULTS - (Stromberg Carburetter)

(1) Faulty running or misfiring - check ignition system, cleanliness and setting of plugs, ignition distributor condenser. Sufficiency of fuel in tank, cleanliness of strainers.

(2) Loss of maximum speed - check that throttle valves can be fully opened, see that power jets are free from obstruction.

(3) Flat spot at small throttle opening - Adjust idling. If flat spot is still evident, examine idle discharge holes and pilot jets for stoppage.

(4) Flat spot at half throttle opening - check main jets and accelerator pump for stoppage.

(5) High fuel consumption - check fuel level in float chamber, see page C8.

NOTE: - During re-assembly, any damage to main body joint, permitting an air leak to vacuum passage (71, Fig.C1), will cause by-pass valve to be open and so allow excess fuel to pass at cruising speeds.

DIS-ASSEMBLY.

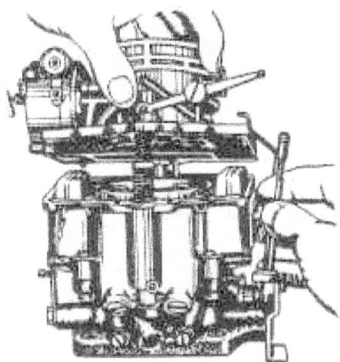

FIG. C3. REMOVING AIR HORN.

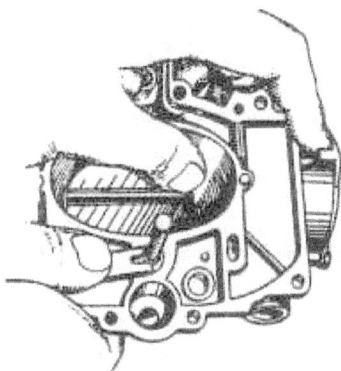

FIG. C4. REMOVING VACUUM POWER PISTON.
(Tool No. T 24773.)

G. C5. REMOVING FLOAT NEEDLE
VALVE SEAT. (Tool No. T 20140.)

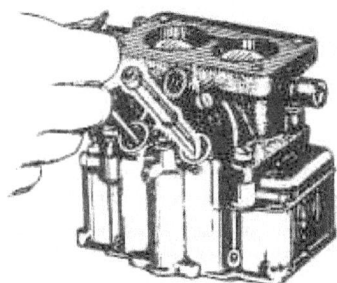

FIG. C6. REMOVING METERING JETS.
(Tool No. T 24924.)

IG. C7. REMOVING MAIN DISCHARGE TUBE
(Tool No. T 24967.)

DIS-ASSEMBLY. (Continued)

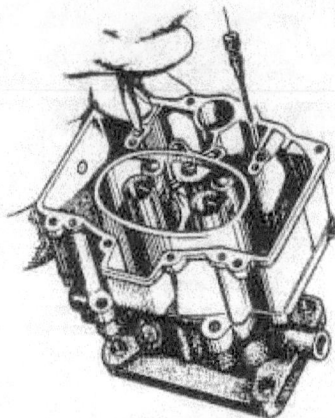

FIG. C8. REMOVING IDLE TUBES.  FIG. C9. REMOVING POWER JETS.

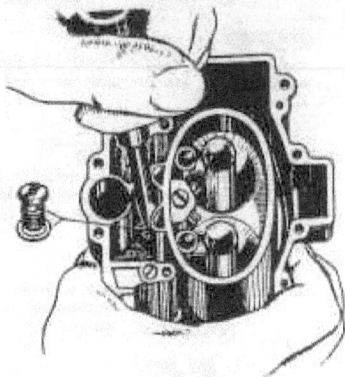

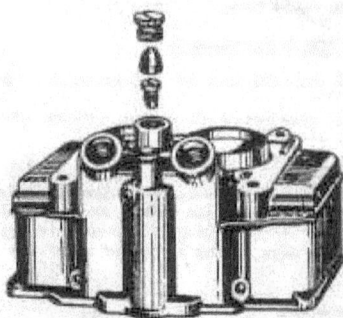

FIG. C10. CHECK VALVE, PLUG & STRAINER.

CARBURETTER DIS-ASSEMBLY (Stromberg).

Remove carburetter from engine.

(1) Remove air silencer.
(2) Disconnect fuel feed pipe.

# SERVICE HANDBOOK

### SILVER WRAITH — SILVER DAWN — BENTLEY MK. V
### R. TYPE BENTLEY — PHANTOM IV.

(3) Disconnect throttle control from lever.
    Disconnect lever from strangler valve spindle.
(4) Remove holding nuts and lift carburetter from manifold.

Dis-assembly:- See Figs. C3 - C10.

## SERVICING.

Gum residue from the fuels may be found deposited on the needle valve assembly and the various jets and orifices. The recommended solvent is denatured alcohol.

### Throttle body.

(1) Examine throttle spindle for wear.
(2) Examine casting for cracks. Remove any carbon or rust from throttle barrels with fine emery cloth.

### Main body.

(1) Examine for cracks, bent flange surfaces or stripped threads.
(2) Inspect main discharge jets for bent tips, and ensure that all channels are clean.

### Air horn.

(1) Inspect strangler valve stem for distortion or wear.
(2) Check floats are not punctured.
(3) Check vacuum power piston bore is not too badly worn for re-use.
(4) Check that kick piston moves freely in housing and that link pin is not too badly worn.

## CARBURETTER RE-ASSEMBLY (Stromberg).

Reverse the procedure for dis-assembly. See Figs. C11 - C15.

To refit carburetter to engine, reverse the dismantling procedure.

## Automatic Starting System.

The two calibrated settings of the automatic starting system are the setting of the thermostat and the actual strangler valve opening by the vacuum piston. This, the "vacuum kick-gap", is measured by inserting gauge of specified diameter between the edge of the strangler valve and the side of the carburetter body.

The "kick-gap" is set to .125" by the manufacturers.

To check thermostat, remove from engine, and allow to obtain room temperature, which should be 70°F. If this temperature is unobtainable, set thermostat one notch rich for every 5° above 70° and, conversely, one notch lean for every 5° below 70°F.

Refer to Fig. C16, unhook thermostat spring from prong (F), loosen locking screw (B) and rotate the indicator point (D) to the "0" (Zero mark E) on the plate (G). In this position, the hook on the thermostat spring should come flush with the prong (F) of the indicator, the lever (A) being held up against the stop. The pointer (D) should then be revolved until it is opposite

SILVER WRAITH — SILVER DAWN — BENTLEY MK. VI.

R. TYPE BENTLEY — PHANTOM IV.

RE-ASSEMBLY.

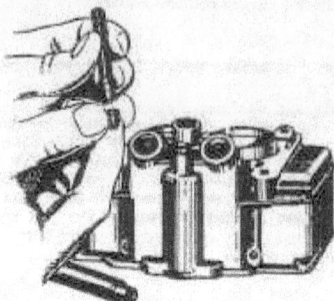

FIG. C11. INSTALLING CHECK VALVE STRAINER (1). (Tool No. T 25097).

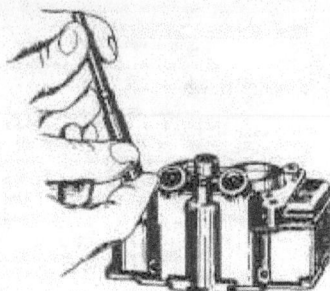

FIG. C12. INSTALLING CHECK VALVE STRAINER (2).

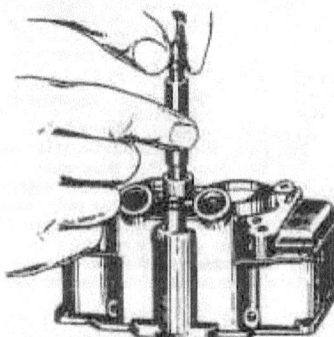

FIG. C13. INSTALLING CHECK VALVE STRAINER (3).

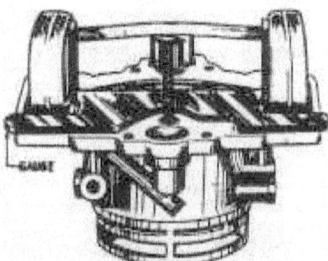

FIG. C14. ASSEMBLY OF ACCELERATOR PUMP NOZZLE.

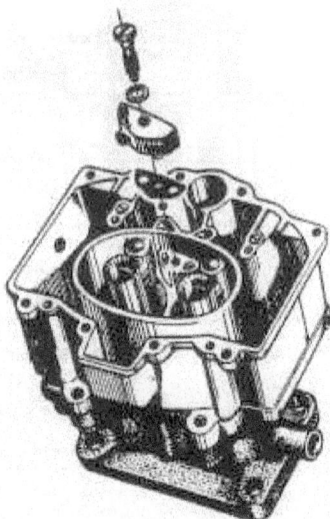

FIG. C15. CHECKING FLOATS. (Tool No. T 24971).

the centre dot (C), stamped on the plate, indicating the correct tension or setting, rich of the zero point. (This dot is 10 notches rich of the zero point.)

Recalibrating Thermostat.

If, after the above check, the thermostat spring hook does not come flush with the prong (F):-

With the lever (A) held up against stop, revolve the pointer cage (D) until the hook on the thermostat spring comes flush with the prong. This will bring pointer to a different position and so provide a fresh location for the zero point. This should then be stamped on the plate and the old mark obliterated. Number off 10 notches on the rich side of the new zero point and mark this with a new centre dot, which will represent the new setting of tension for the thermostat.

The rod connecting the thermostat to the strangler valve, should be so adjusted for length, that when the strangler valve is fully closed, the lever (A) on the thermostat is approximately .050" off the stop pin (B).

Checking Fuel Level in Float Chamber.

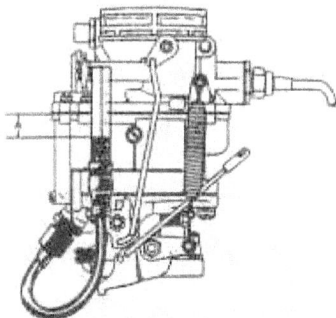

FIG. C16. THERMOSTAT.

The fuel level should be 9/16" below the top face of the main body, and is best checked by connecting sighting tube after removing the main discharge jet plug, as illustrated in Fig. C17. A sighting plug is provided on the side of the carburetter main body but this is a less accurate method of checking.

FIG. C17. FUEL LEVEL CHECK.

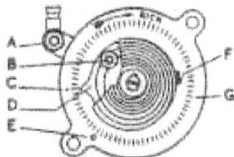

ZENITH, TYPE D.B.V.C.42 CARBURETTER.

JET DATA.

| | A.C. Air Cleaner. | | | Oil Bath Cleaner. | | |
|---|---|---|---|---|---|---|
| | Sea Level to 3,000 ft. | 3,000 ft. to 5,000 ft. | 5,000 ft. to 8,000 ft. | Sea Level to 3,000 ft. | 3,000 ft. to 5,000 ft. | 5,000 ft. to 8,000 ft. |
| Venturi | 1¼" | 1¼" | 1¼" | 1¼" | 1¼" | 1¼" |
| Main Jet | .060" | .058" | .056" | .058" | .056" | .054" |
| By-pass Jet | .054" | .054" | .054" | .050" | .050" | .050" |
| Pump Jet | No. 70 | No. 70 | No. 70 | No. 70 | No. 70 | No. 70 |
| Vacuum Piston | .207"/.202" | | | .207"/.202" | | |
| Setting | .781"/.718" | | | .781"/.718" | | |
| Petrol Level | Below hole face. | | | Below hole face. | | |
| 9' Head | | | | | | |
| Thermostat | 12 notches rich. | | | 12 notches rich. | | |

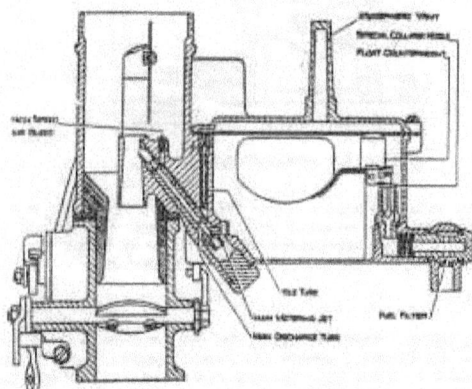

FIG. C18. CARBURETTER SECTION.

ZENITH CARBURETTER ADJUSTMENT.

The carburetter is carefully adjusted to the engine before leaving the factory and the settings should not be altered unnecessarily.

There are two adjustments on the carburetter, one for idling speed and one for the idling mixture. Both of these should be made together, and only when the engine is well warmed up.

(1) The throttle stop screw should be set so that the engine runs at approximately 350 r.p.m.

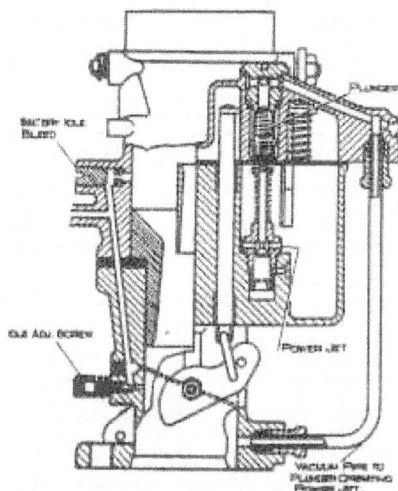

FIG. C19.  CARBURETTER SECTION.

(2)  Adjust the idle adjustment screw (Fig.C19) to give the most even running.
Turning the screw IN weakens the mixture, and turning the screw OUT
enrichens it.   The approximate position is 1½ turns back from the
closed position.   Operate screw with fingers only.

DIAGNOSIS OF FAULTS.

For loss of maximum speed - Check throttle for full opening, check by-pass jet for
stoppage.   In order to examine, remove the float chamber cover and screw out the
by-pass valve complete.   Check also by-pass vacuum piston for sticking in the
"UP" position.   Examine main jets for stoppage.

Flat spot at small throttle opening - Adjust idling to give more regular engine
rhythm.   If flat spot is still evident, examine idle discharge holes and idle
tube for stoppage.

Flat spot at half throttle - Examine main jet for stoppage.   Check accelerator
pump for stoppage.

High fuel consumption - Check jets and float level for setting, which should be ½"
from float cover facing.   Level should be checked with the engine idling and
vehicle on level ground.   If setting correct, check for sticking or leaking by-
pass valve, which will allow flow of fuel at part throttle.   Check economiser
valve for sticking, also, clean and check ball valve at base of pump cylinder, as
if not seating, petrol will issue from pump jet at all times.   Check tightness of
vacuum piston plug on float chamber cover.

SILVER WRAITH — SILVER DAWN — BENTLEY MK. VI.

R. TYPE BENTLEY — PHANTOM IV.

Note: Important:- The external vacuum connection unions must be tight, as leakage at this point, with the consequent partial loss of vacuum, will cause the by-pass valve to open too early.

## DIS-ASSEMBLY.

### Removal from Engine.

(i) Remove the air silencer.
(ii) Disconnect the fuel feed pipe from the union.
(iii) Disconnect the throttle control lever from the cam assembly.
(iv) Disconnect the strangler control rod.
(v) Remove the two retaining nuts and lift off the carburetter.
(vi) Remove retaining nuts and withdraw thermostat from housing in manifold.

### Dismantling Carburetter Body.

(i) Disconnect the vacuum pipe union from the float chamber cover.

(ii) Using a 7/16" spanner, loosen the economy vacuum piston plug.

(iii) Remove the three setscrews and take off the float chamber cover. If in good condition, the gasket should be preserved; if not, it must be replaced.

(iv) Remove the economy vacuum piston assembly from the float chamber cover (Fig.C20).

(v) Unscrew the float retaining pin, remove float and withdraw the

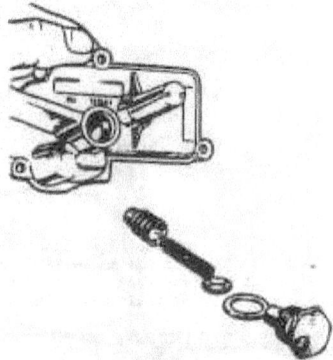

FIG.C20.   ECONOMY PISTON ASSEMBLY.

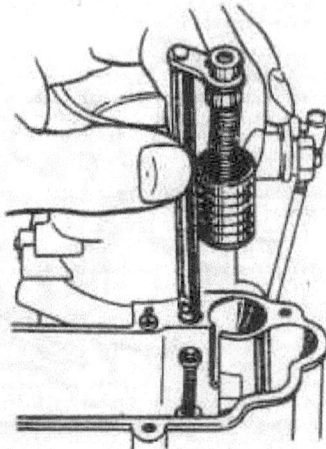

needle valve.
(vi) Using a $\frac{1}{4}$" box spanner, remove the needle valve seating.
(vii) Having previously disconnected the link, lift out the accelerator pump (Fig.C21).
(viii) Using a suitable box spanner, remove the by-pass valve (Fig.C22).
(ix) Using a box spanner, remove the pump feed valve.
(x) Using a box spanner, remove the pump discharge ball valve (Fig.C23).

FIG.C21.   ACCELERATOR PUMP.

# SERVICE HANDBOOK

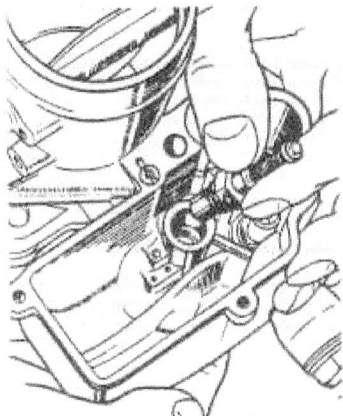

FIG.C22.   THE BY-PASS VALVE.

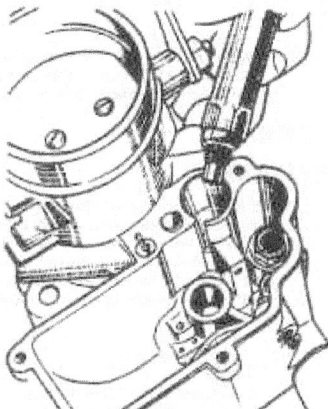

FIG.C.23.   PUMP DISCHARGE BALL VALVE

(xi)  Unscrew the pump jet plug and remove the jet.

(xii)  Unscrew and remove the idle tube (Fig.C24).

(xiii) Remove the main jet plug and washer with a 5/16" spanner.

(xiv)  Using the special key, withdraw the metering jet.

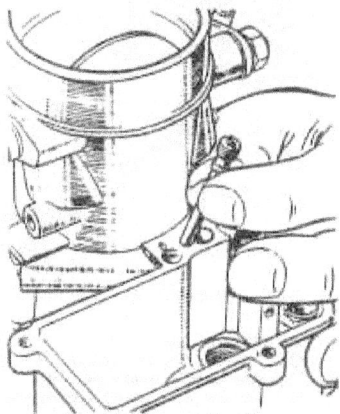

FIG. C.24.  THE IDLE TUBE.

xv) Using the special tool, withdraw the main discharge jet.  Unless the special tool is available, no attempt must be made to remove the jet, never attempt to drive it out.  It is, however, rarely necessary to remove the jet, and should be undertaken only in cases of blockage or damage (Fig.C25.)

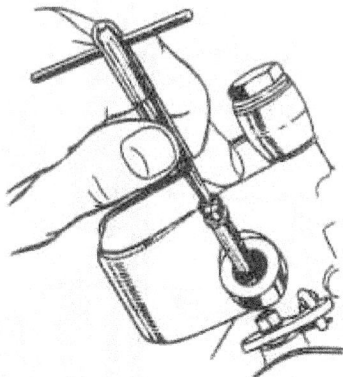

FIG.C25. THE MAIN DISCHARGE JET.

- C.12 -

SILVER WRAITH — SILVER DAWN — BENTLEY MK. VI.

R. TYPE BENTLEY — PHANTOM IV.

(xvi)  The high speed bleed may be removed and replaced, after unscrewing, with a small greased screwdriver.

(xvii)  Unscrew and withdraw the idle discharge plugs.

(xviii)  Separate the throttle and main bodies, then remove the large venturi. Care must be taken that the thick washer is undamaged. The cam rod should be disconnected and it should be specially noted that the correct way to hold, is to support the strangler lever with the fore-finger, thus preventing damage to the pick-up pin in the cam assembly.

## Dismantling Strangler.

(i)  Holding the strangler lever in the manner already indicated, remove the cam lever rod from the valve lever.

(ii)  Remove the two securing screws in the strangler butterfly valve and remove it from the spindle.

(iii)  Remove pinch bolt screw, spring, distance piece and plastic washer, taking care not to damage the plastic washer.

(iv)  Remove the spindle. It should be noted that the cast iron bushes inserted in the die casting are not renewable.

## Inspection and Cleaning.

Inspect and clean all parts thoroughly. Gum deposits from the petrol may be removed by soaking in de-natured alcohol and then scrubbing with a stiff bristled brush.

When cleaning, on no account pass a wire or hard material through the jet orifices. These orifices have been carefully calibrated, and the slightest alteration of shape or size may interfere with the efficiency of the carburetter; use a stiff bristled brush and compressed air.

## RE-ASSEMBLING.

Re-assembling the carburetter is the reverse of the dismantling instructions, except for the following points:-

(i)  Ensure that all washers are in good and undamaged condition.

(ii)  The main discharge jet must be replaced with the groove as indicated in Fig.C25.

(iii)  The vacuum pipe must be re-attached to the float chamber cover before the cover is secured with the three screws.

## The Strangler.

(i)  Replace the spindle, screw the butterfly into position, but do not tighten the screws.

(ii)  Slacken the pinch bolt on the strangler lever and ease it a little towards the end of the spindle.

(iii)  Replace the plastic washer, distance piece and spring at the other end of the spindle.

(iv)  Adjust the position of the strangler lever on the spindle by tapping gently to obtain the correct end play (i.e. .003" - .006"), then tighten the clamping bolt.

(v)  Hold the strangler valve in the closed position to centralise it, and tighten the two securing screws.

(vi)  Re-attach the cam lever rod, supporting it firmly (as already described) when tightening, to avoid damage to the cam assembly.

(vii)  Finally, test that the movement is free.

# SERVICE HANDBOOK

## Checking and Adjusting the Automatic Choke.

The adjustment of the "Automatic Choke" portion of the carburetter consists of six mechanical settings. No other adjustment on the engine is necessary.

These adjustments consist of the following:-

1. Thermostat setting.

2. Thermostat lever and setting pin.

3. "Vacuum kick" gap.

4,5. The position of the fast idle stop screw on the cam with the strangler (a) closed for cold starting, and (b) in the "vacuum kick" position.

6. The fast idle throttle gap.

## 1. The Thermostat Setting.

The thermostat spring is sensitive to variation of temperature, therefore the setting check must be made in a room with a temperature of 70°F. The thermostat should be left for an hour before setting, in the room, to adjust itself to this temperature.

(i)    Loosen nut R (Fig.C.26).
(ii)   Unhook the end of the spring from the retaining prong S.
(iii)  With the spring free, hold the lever T against the check pin U in the closed position. Note that the figure shows a "Right-hand" thermostat.
(iv)   Rotate the prong S, until it just touches the <u>outside</u> of the spring loop.
(v)    Check that the pointer V is now opposite "O" mark.
(vi)   As a result of mishandling, it may be necessary to remove the old "O" mark and to make a new one opposite the pointer.
(vii)  Check and reposition, if necessary, the position of the centre dot marking the tension setting.
       If the zero point has been altered, so also must the setting point be altered. The setting must be made to the specification which is given as a number of notches rich (see page 9).
(viii) Slip the loop of the spring over the prong S.
(ix)   Turn the cage until the pointer V is opposite the centre dot.
(x)    Tighten the nut R.

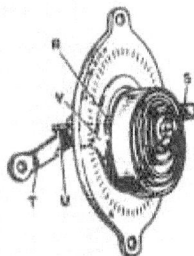

FIG.C26. THE THERMOSTAT.

## 2. The Thermostat Lever and Setting Pin.

(i)    Check that the thermostat rod ball joint is engaged in the correct hole in the strangler spindle lever, i.e. the centre hole.
(ii)   Hold the strangler valve in the fully closed position.
(iii)  Check the gap between the thermostat lever and the setting pin with a drill shank of the specified size, i.e. .09" - .12", 42/43 drill size (Fig.C27).
(iv)   If the gap is incorrect, adjust the ball joint thread to the correct gap size. The ball joint must be disconnected from the strangler, before making this adjustment.

SILVER WRAITH — SILVER DAWN — BENTLEY MK. VI.

R. TYPE BENTLEY — PHANTOM IV.

3. Checking the "Vacuum Kick" Gap.

(i)    Remove the grub screw.

(ii)   Insert a piece of stiff wire into the grub screw hole, lightly holding the strangler towards the closed position, press the vacuum piston to its stop screw.

(iii)  Check that the fast idle now rests on the lowest step of the cam.

(iv)  Check the vacuum kick gap between the top of the strangler valve and the body with a drill of the specified diameter (see Fig.C28.).

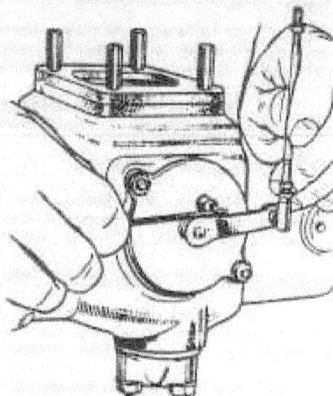

4,5. Adjust the Fast Idle Stop Screw on the Cam and the "Vacuum Kick" Gap.

(See also "The Fast Idle Throttle Gap", page C18.)

FIG. C.27. CHECKING GAP - THERMOSTAT LEVER AND CHECKING PIN.

It should be noted that, with the strangler fully closed, the fast idle stop should rest on the middle step of the cam.

With the strangler opened to the "Vacuum Kick" position, the stop screw should rest on the lowest step of the cam.

(i)    If the "Vacuum Kick" gap is correct - The position of the fast idle stop screw on the cam may be adjusted by bending the stop tag on the cam lever, as required, with a pair of pliers or a screwdriver (Fig.C.30).
Remember that when the strangler valve is opened to the "Vacuum Kick" position, the stop screw must rest on the lower step.

(ii)   If the "Vacuum Kick" gap is incorrect - Try to adjust it without disturbing the vacuum kick piston stop pin. If possible, adjust the gap by twisting the left and right-hand threaded cam rod with a pair of pliers.

Then check the position of the fast idle stop screw and, if necessary, adjust as described previously. If adjustment, as described in 3 No.(ii), is found to be impossible, then the vacuum piston assembly must be removed.

Dismantling Vacuum Piston Assembly.

(a)   Remove the two screws securing the cam plate.

(b)   Swing the cam assembly clear without unlinking the cam rod. This must be done before the vacuum piston assembly is unscrewed, or the pick-up lever will foul the piston and prevent it being turned.

(c)   Using a large broad ended screwdriver, unscrew the vacuum cylinder and piston of the carburetter.

(d)   Separate the piston from the cylinder (Fig.C29.).

(e)   Unscrew the piston stop pin with a pair of long-nosed pliers.

(f)   Clear the lead sealing plug from the end of the pin and interior screw threads.

Dismantling the Cam Assembly.

      If it should be found necessary to dismantle the cam assembly, this can be done easily by unscrewing the nut and withdrawing the stem.   Check the pick-up pin for distortion due to maladjustment.

      It is essential to observe the sequence of dismantling and the relative position of each part, to ensure correct re-assembly.

Re-assembling the Piston.

      Replace the vacuum piston assembly, ensuring that the thin copper washer is in place under the head of the cylinder.

Re-Assembling and Replacing the Cam.

      In re-assembling the cam, care must be taken when attaching the cam spring. The method of attachment and assembly are:

  (a) Replace the cam assembly, ensuring that the pick-up pin is correctly positioned.   Care must also be taken to ensure that the pick-up lever is engaged in the slot of the piston.

  (b) Replace the two securing set-screws in the cam plate.

  (c) Adjust the vacuum kick piston stop pin until the correct vacuum kick gap is obtained (Fig.C28).

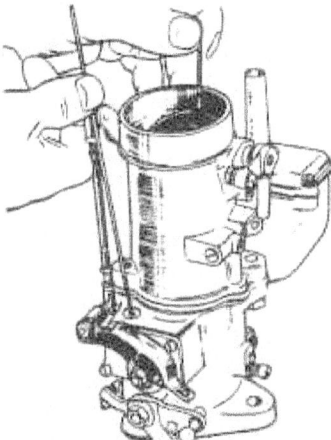

FIG. C28.   CHECKING VACUUM KICK GAP.

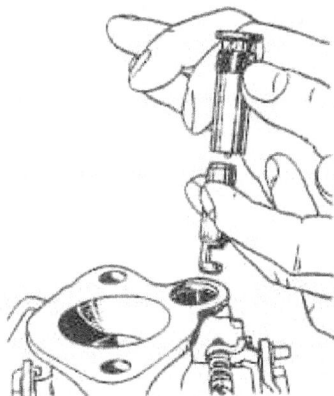

FIG.C29.   VACUUM PISTON & CYLINDER.

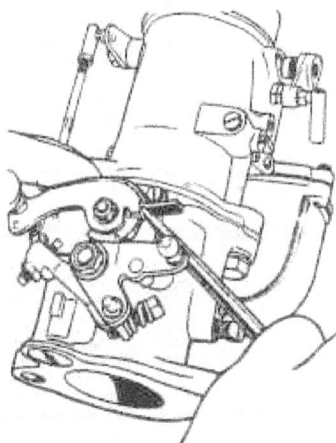

FIG.C30.   ADJUSTING CAM LEVER STOP TAG.

- C.16 -

# SERVICE HANDBOOK

### SILVER WRAITH — SILVER DAWN — BENTLEY MK. VI.
### R. TYPE BENTLEY — PHANTOM IV.

(d) Invert the carburetter, and drop a lead shot into the stop pin hole. Seal with a sharp blow on a thin punch, this locks the stop pin in position and seals it.

6.  The Fast Idle Throttle Gap.

(i)   With the strangler valve held in its fully closed position, check that the fast idle screw rests on the centre step of the cam.

(ii)  Check that the throttle valve is open sufficiently to pass a drill .031" - .033" dia., between the butterfly and the throttle body. It should be noted that the gap must be gauged at its widest point, i.e. at the idle discharge holes.

(iii) If necessary, adjust the gap by the fast idle stop screw.

S.U. CARBURETTERS - MANUAL CONTROLLED TYPE.

JET DATA.

S.U. Carburetter - H.4, 1¼" dia. choke, 3.500" bore engine.   Carburetter has 2 bolt fixing.

Jet size    - .100".
Jet needle  - L.B.1 Bentley air cleaner.
            - S.O.  A.C. air cleaner, commencing chassis B.2.BH.
            - S.F.  Oil bath air cleaner.

S.U. Carburetter - H.6, 1¾" dia. choke, 3.500" bore engine.   Carburetter has 4 bolt fixing.   Commencing Chassis B.83.HP.

Jet size    - .100".
Jet needle  - S.J.  A.C. air cleaner.
            - S.F.  Oil bath air cleaner.

S.U. Carburetter - H.6, 1¾" dia. choke, 3.625" bore engine.   Carburette has 4 bolt fixing.

Jet size    - .100".
Jet needle  - S.F.  A.C. air cleaner.

ADJUSTMENT - S.U. CARBURETTERS.

There are only two adjustments on the carburetter, the jet adjusting nut for the mixture, and the idle adjusting screw for the idling speed.   Both these should be carried out together and with the engine well warmed up.

(1) <u>Mixture</u> - the jet adjusting nut acts as a "stop" against which the jet head should bear, except when jet is lowered for cold starting. The adjustment consists of varying the position of this "stop" and so varying the amount by which the jet is raised for normal running.

If the engine has a constant uneven beat (hunting), this is due to a rich mixture. If the exhaust note is irregular and splashy, the mixture is too weak.

(2) <u>Throttle</u> - the correct idling speed is obtained by adjusting the throttle stop screw.

As two carburetters are fitted, correct synchronisation is required.

<u>Synchronising throttles</u> - This is mostly an aural check and largely a matter of trial and error. The air intake must be removed to "listen in" to the sound of the air flow through the carburetters.

(1) Free the connecting shaft and fit a temporary throttle stop to the rear carburetter, and with the engine well warmed up and running at a fast tick-over (500 - 600 r.p.m.), adjust each carburetter by means of the throttle stop screws, "aurally" and for "equal depression" of the pistons.

(2) Slightly advance the opening of the rear carburetter and nip up the pinch bolt on the connecting shaft. Re-check and re-adjust if necessary to correct. Afterwards, remove rear throttle stop screw and adjust as necessary on front carburetter for correct idling.

<u>SERVICING. - S.U. Carburetters.</u>

Gum residue from the fuels may cause sticking of needle or piston, this should be removed by cleaning with a solvent, such as denatured alcohol.

<u>Hydraulic Suction Piston Damper.</u>

This device, located in the hollow piston rod, is attached to the oil cap nut; it consists of a plunger with a one-way valve and its function is to give a slightly enrichened mixture by preventing the piston from rising unduly quickly on acceleration. The only attention necessary is to keep it supplied with thin oil.

<u>Centring of Jet</u> - S.U. Carburetter.

Should it be essential to remove the jet, this can be done by unscrewing the jet holding screw. It must be understood that the needle is very nearly as large as the jet, and yet must not touch it. Therefore, careful assembly is needed when centring jet to needle.

First disconnect jet head from jet operating lever. Withdraw jet completely and remove adjusting nut and spring. Replace nut without spring and screw up to highest position. Next, feel that piston is perfectly free by lifting with finger. If not, slacken jet screw and manipulate lower part of assembly, including the projecting part of the bottom half jet bearing, adjusting nut and jet head. This assembly should now be slightly loose. The piston should now rise and fall quite freely as the needle is now able to move the jet into the required central position. Tighten the jet screw and check that piston is now quite free, if not, slacken off and repeat operation. When complete freedom of piston is achieved, remove jet adjusting nut and jet, replace spring and screw back to its original position.

SILVER WRAITH — SILVER DAWN — BENTLEY MK. VI.

R. TYPE BENTLEY — PHANTOM IV.

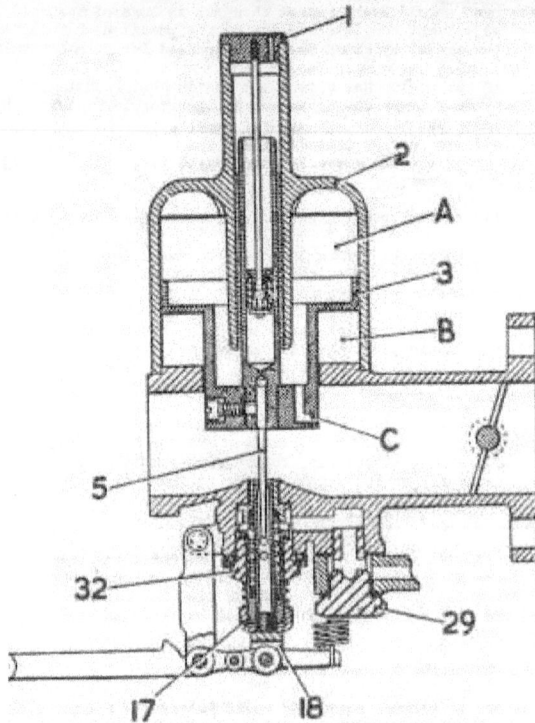

FIG. C31.  DIAGRAMMATIC SECTION - S.U. CARBURETTER.

  1.  Hydraulic damper rod and cap.
  2.  Automatic air valve cylinder.
  3.  Suction disc.
  5.  Jet needle.
 17.  Jet adjusting nut.
 18.  Jet head.
 29.  Float chamber retaining bolt.
 32.  Jet holding screw.
  A.  Upper side of piston (Under depression).
  B.  Lower side of piston (To atmosphere).
  C.  Connecting channel to A.

## Checking Fuel Level in Float Chamber.

The fuel level in the float chamber can be varied by bending the forked lever (43).

The forked lever should be set so that when it is holding the needle against its seating, a 7/16" dia. rod can just be passed between the lever and the float chamber cover as illustrated in Fig.C.32.

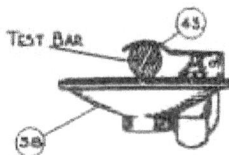

FIG. C.32. CHECKING FLOAT LEVEL.

38. Float chamber cover.
43. Float lever.

## S.U. CARBURETTERS - AUTOMATIC TYPE.

### JET DATA.

Choke size  -  1¾".
Jet   size  -  .100".
Jet needle  -  S.H.

The general working principles of the automatic type carburetter are similar to those as previously described for the non-automatic type.  The main difference being that the hand throttle and also the mixture control has been eliminated, and that an electro-thermostatic control has been incorporated for starting purposes.

The automatic system consists of:-

(1)  An out of balance butterfly valve in the air intake, indirectly coupled to a "kick diaphragm" located in a  housing on the side of the air intake. This diaphragm is subject to induction depression by means of a connecting pipe from the diaphragm housing to the induction manifold.

(2)  An electro magnet wired in parallel in the starter relay circuit, which holds the butterfly in the closed position for cold starting.

(3)  A fast idle cam connected by a rod to the butterfly spindle.

(4)  A thermostatic coil coupled to the butterfly valve and subject to engine coolant temperature.

## OPERATION.

Before attempting to start, the accelerator pedal must be depressed in order to release the fast idle cam and allow the thermostatic spring to close the butterfly.  On releasing the pedal, the throttle opening becomes greater than for normal idling, as the extra stop is resting on the fast idling cam.  In cases of starting under low ambient temperature conditions, the accelerator should be held depressed, about half of normal travel, whilst the starter button is operated.

- C.20 -

### SILVER WRAITH — SILVER DAWN — BENTLEY MK. VI.
### R. TYPE BENTLEY — PHANTOM IV.

Pressing the starter button causes the electro-magnet to hold the butterfly in the closed position. When the engine fires, the button is released, the magnet ceases to hold the butterfly closed and the induction pipe depression operates on the kick diaphragm to move the loose lever on the butterfly spindle, the lever opening the butterfly to a predetermined amount.

As the engine warms up, the thermostatic spring loses tension and the butterfly opens; but prior to this, the depression, acting on the kick diaphragm, is such that the diaphragm can hold the butterfly open against the thermostatic spring for the drive away. Further depressing the accelerator pedal in driving away, causes the fast idle stop to move away from the cam; the air flow passing the off-set butterfly, assisting in opening for any engine requirements more than a fast idle.

Certain chassis in RT, RS and SR series may suffer from flooding of carburetter when starting from cold, an over-ride has been incorporated on later productions whereby, when the accelerator pedal is fully depressed, the choke valve is mechanically opened by trip levers. Requisite parts for this modification are available from the London Service Station.

ADJUSTMENT:

(Before installing carburetter in position.)

(1) With the butterfly closed, a radial clearance of .010" is required between the pin (1, Fig.C33) and the pick-up lever (2), adjust by means of washers placed on kick diaphragm spindle.

(2) Fit the diaphragm assembly and adjust the kick-gap to .090", see Fig.C34.

(3) With the butterfly closed, adjust the air gap between the lever (1) and the solenoid yoke, to .002"/ .006" by means of shims (Fig.C35). It is essential that the faces are parallel, to ensure magnetic flux density.

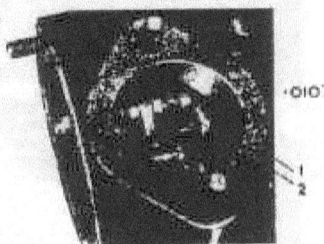

FIG.C33. PICK-UP LEVER CLEARANCE.

1. Lever.
2. Pin.

CHOKE VALVE ASSEMBLY COVER

SCREW CONTROLLING TRAVEL OF KICK DIAPHRAM

.090" DRILL

FIG.C34. KICK DIAPHRAGM ADJUSTMENT.

### SILVER WRAITH — SILVER DAWN — BENTLEY MK. VI.
### R. TYPE BENTLEY — PHANTOM IV.

(With the carburetters installed on engine.)

1) Fit butterfly body complete to air intake and connect depression pipe between kick diaphragm and induction pipe.

2) With front carburetter throttle fully closed, tighten the coupling pinch bolt to give .098" clearance between the fast idle cam and the cam lever (Fig.C36); the cam adjusting screw hole should be central with cam face. Hold the rear carburetter throttle closed and tighten the rear coupling pinch bolt.

FIG.C35. SOLENOID YOKE ADJUSTMENT.

(3) Fit thermostat unit.

(4) Screw in throttle stop ½ turn. Fit screw (1) and locknut to lever (1A, Fig. C.37). With screw resting on fast idle cam (3), adjust screw to give .030" clearance between throttle stop screw and stop lug.

(5) With butterfly closed, adjust rod (1, Fig.C38) to give 1/32" clearance between thermostat lever and stop (2 and 3). Adjust fast idle pick-up rods to butterfly.

THROTTLE FULLY CLOSED

IG.C36. FAST IDLE CAM AND LEVER ADJUSTMENT.

1. Fast Idle Adjusting Screw.
2. Fast Idle Cam Stop.
3. Fast Idle Cam.
4. Fast Idle Cam Link.
5. Rod, Butterfly to Fast Idle Cam.
6. Rod, Thermostat to Butterfly.

FIG.C37. THROTTLE STOP ADJUSTMENT.

FINAL TEST ADJUSTMENT.

To alter jet adjustment, remove cap nut from carburetter jet, and rotate jet holding screw, upwards to weaken mixture and downwards to strengthen.  With the engine warm, finally set the mixture for idling.

FIG.C38.  THERMOSTAT LEVER ADJUSTMENT.

Allow the engine to cool, and readjust the fast idle cam screw (1, Fig.C37) to obtain .030" clearance between throttle stop and lug.  Ensure consistent setting as a difference of .002"/.003" makes a difference to starting speed of engine.  Adjustment of the fast idle cam pick-up rod (5) assists during the warming up period, the length of the rod alters the relative positions of the fast idle screw to the fast idle cam.

THERMOSTAT.

Any checking and resetting of the thermostat spring must be done at a room temperature of 68°F, and the unit should be left for two hours in this temperature to stabilise.

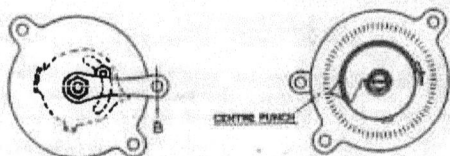

FIG.C39.  BI-METAL COIL ADJUSTMENT.

The coils are pre-set to a loading of 200 - 205 grammes.  This weight should just move the lever clear of the stop in the direction of arrow 'B', Fig.C39.

Since the coils are temperature sensitive, they must not be handled whilst adjusting.

### SILVER WRAITH — SILVER DAWN — BENTLEY MK. VI.
### R. TYPE BENTLEY — PHANTOM IV.

FUEL FILTERS.

The main fuel filter mounted at the rear of the frame and close to the outlet from the fuel tank, is provided with two circular gauzes located above a settling sump, Fig.C40.   Fuel passes upwards through these gauzes and dirt settles on their lower faces and in the sump.

Service by dis-assembling and cleaning gauzes in petrol, and clean out sump.

A small gauze filter is also fitted in the inlet union to each carburetter.   These should be periodically removed and serviced.

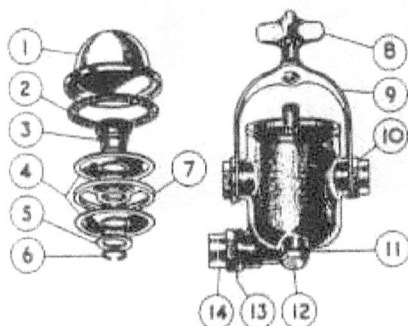

1. Cover.
2. Cork washer.
3. Knurled nut.
4. Gauzes.
5. Washer.
6. Spring clip.
7. Distance piece.
8. Wing nut.
9. Stirrup.
10. Inlet union.
11. Aluminium washer.
12. Drain plug.
13. Aluminium washer.
14. Outlet union.

FIG.C40.   MAIN FUEL FILTER.

FUEL PUMPS.

Twin fuel pump units are fitted, each pump works independently of the other, but both deliver fuel to a common chamber.

Servicing.

(1)   Restriction in System on Delivery Side - if pump functions, check carburetter float chamber needle for sticking.

(2)   Restriction in System on Suction Side - disconnect inlet pipe line, if pump functions, check rear strainer and pipe line.   Also, check filter in pump body.   Restriction on suction side will cause pump to become overheated.

(3)   Failure of Electric Supply - if pump does not function, check electric supply by connecting 12 volt bulb between supply lead and pump body.

(4)   Pump Valves Stuck to Seating - Unlikely unless car has been laid up for considerable time, remove valves, clean and replace, smooth sides downwards.

(5)   Faulty Pumping Unit - disconnect and check points and flexible connections.   Clean points if necessary.

SILVER WRAITH — SILVER DAWN — BENTLEY MK. VI.

R. TYPE BENTLEY — PHANTOM IV.

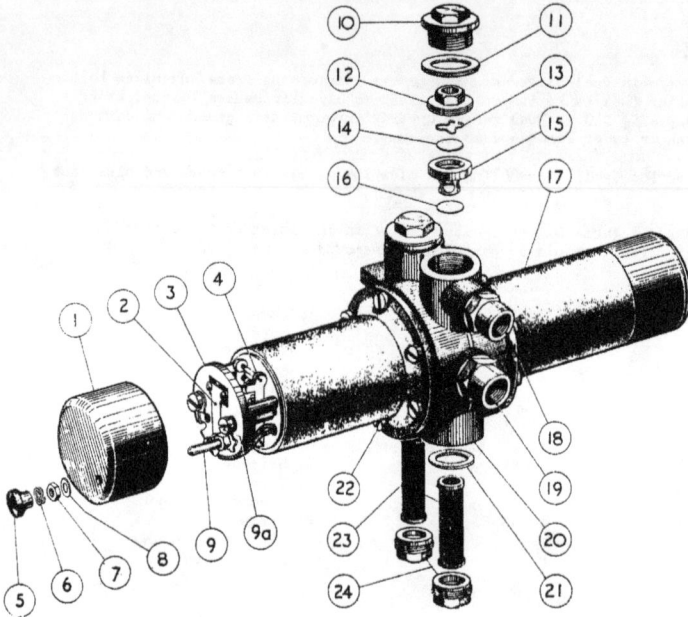

FIG.C41. FUEL PUMPS.

|   |   |   |   |
|---|---|---|---|
| 1. | End cover. | 14. | Delivery valve disc. |
| 2. | Spring blade. | 15. | Delivery valve cage. |
| 3. | Bakelite pedestal. | 16. | Inlet valve disc. |
| 4. | Outer rocker. | 17. | Magnet housing. |
| 5. | Terminal nut. | 18. | Outlet union. |
| 6,7 & 8. | Washers and nut. | 19. | Inlet union. |
| 9. | Terminal Screw. | 20. | Body. |
| 9A. | Fixing Screw. | 21. | Fibre washer. |
| 10. | Cap-nut. | 22. | Fixing screw. |
| 11. | C and A washer. | 23. | Filter. |
| 12. | Retaining ring. | 24. | Filter plug. |
| 13. | Clip, valve disc. |   |   |

(6)  Diaphragm Swollen or Armature Jammed - if points are apart and will
not make contact, or pump attempt suction stroke but fails to make points
separate, remove diaphragm assembly for inspection. Also, check fibre
bushes on outer rockers, failure of outer rockers to "throw-over" may be
due to swollen fibre bushes.

## SILVER WRAITH — SILVER DAWN — BENTLEY MK. VI.
## R. TYPE BENTLEY — PHANTOM IV.

(a) <u>To remove diaphragm assembly</u>, remove magnet housing complete with diaphragm assembly. Unstick diaphragm from flange of housing and unscrew anti-clockwise. Remove brass rollers and check for "flats" on spherical edges, renew if worn.

(b) <u>To re-assemble and adjust diaphragm</u>, slacken off spring blade retaining screw so that no pressure is exerted on tungsten points on outer rockers. (If this is not done, diaphragm cannot be correctly set.)

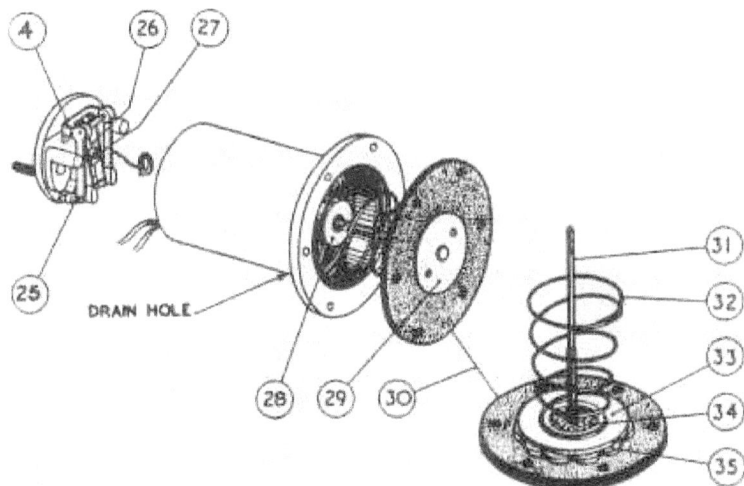

FIG.C42.  CONTACT MECHANISM.

|  |  |  |  |
|---|---|---|---|
| 4. | Outer rocker. | 30. | Diaphragm assembly. |
| 25. | Hinge pin. | 31. | Armature rod. |
| 26. | Inner rocker. | 32. | Armature spring. |
| 27. | Trunnion. | 33. | Armature. |
| 28. | Magnet core. | 34. | Impact washer. |
| 29. | Brass plate. | 35. | Rollers. |

Re-assemble parts into magnet housing, and holding housing as shown in Fig.C43, unscrew the armature one sixth of a turn (one flange hole) at a time and simultaneously press in and out, until a point is reached, at which the outer rocker will "toggle over" (when the diaphragm is pressed in), then unscrew it a further two thirds of a turn. The setting is now correct. Re-tighten spring blade screw.

NOTE:- Keep the spring blade out of contact and press firmly and steadily on the diaphragm assembly while setting it.

**SILVER WRAITH — SILVER DAWN — BENTLEY MK. VI.**

**R. TYPE BENTLEY — PHANTOM IV.**

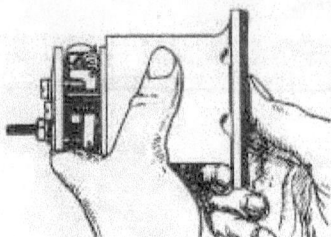

FIG.C43.  ADJUSTING DIAPHRAGM.

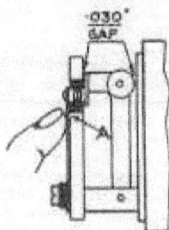

FIG.C.44.  CHECKING ROCKER CLEARANCE.

Stretch the diaphragm to the end of its stroke while tightening retaining screw.

Do not attempt to move core of magnet under any conditions.

Do not stretch the armature spring.

(7)  To Cure a Noisy Pump - if the pump ticks excessively, check for air leak on suction side, from tank to pump.  If pump continues to tick without delivering petrol, check for foreign matter under valves.

AIR CLEANERS.

FIG.C45.  "BENTLEY" TYPE AIR CLEANER.

FIG.C46.  A.C. AIR CLEANER.

The standard air cleaner is the A.C. copper mesh type as illustrated in Fig.C.46.  Early type Bentley Mk.VI chassis were fitted with a Bentley manufactured similar type as in Fig.C.45.

The gauze mesh should be periodically removed and washed in petrol paraffin, dipped in engine oil, drained, and then replaced.

SILVER WRAITH — SILVER DAWN — BENTLEY MK. VI.
R. TYPE BENTLEY — PHANTOM IV.

For use overseas, an Oil Bath type is available, see Fig. C47. As this type of cleaner gathers a considerable amount of dust and dirt, frequent servicing is necessary.

Unscrew nut from top, and lower filter bowl; discard oil and remove filter element. Wash element and bowl in petrol, replace element, fill in oil to correct level and refix.

1. Retaining screw.
2. Oil bowl.
3. Element.

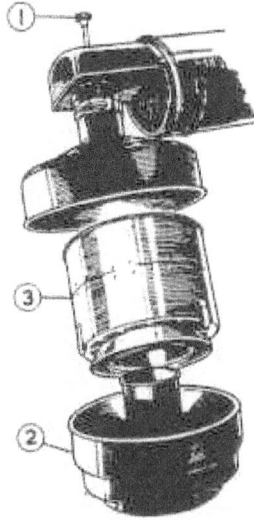

FIG.C47. OIL BATH AIR CLEANER.

# SERVICE HANDBOOK

### SILVER WRAITH — SILVER DAWN — BENTLEY MK. VI.
### R. TYPE BENTLEY — PHANTOM IV.

## THE MODIFIED AUTOMATIC S.U. CARBURETTER

### OIL PRESSURE SOLENOID SWITCH.

This switch has been introduced to ensure more positive starting from cold by delaying the opening of the choke valve. With the earlier system the choke was held fully closed by a solenoid wired in parallel with the starter button. When the engine fired and the starter button was released, the solenoid at the same time released the choke valve, which opened, due to engine depression and sometimes oscilated before achieving the position determined by the kick diaphragm travel, thereby causing stalling.

The break in the solenoid circuit is now delayed by 5 to 7 seconds after the engine has started by wiring the solenoid in parallel with the ignition circuit and the introduction of an oil pressure switch to break the circuit at a predetermined oil pressure.

To prevent choking the solenoid lever is spring loaded to the choke spindle, allowing the manifold depression to open the choke valve against the spring loading while the solenoid is in operation. The spring is adjusted to give a maximum choke opening of 5° thereby ensuring starting and running.

On reaching an oil pressure of 15 lbs/sq.in. the switch breaks the solenoid circuit and the choke is opened another 5° by the kick diaphragm operated by manifold depression. The oil pressure switch is fitted between the pump and the filter, and the solenoid now fitted is a two pole unit, but although energised whenever the oil pressure falls below 15 lbs/sq.in. it is only magnetically strong enough to close the choke when the valve is within 5° of the closed position.

In the event of flooding, the choke valve spindle pin is allowed 30° of free travel in the solenoid lever, therefore the choke can be opened 30° by fully depressing the accelerator pedal even though the solenoid is in operation.

### Settings.

(1) Kick Diaphragm Travel.

This should be set .075" by means of a No.46 drill between the choke valve and the body.

(2) Solenoid Air Gap.

The solenoid air gap is set to between .0005" and .004" and adjusted by shims under the solenoid flange.

(3) Spring Tension Setting.

The spring tension should be set so that a weight of 235 to 240 grammes acting on a 2" arm just open the valve sufficiently to allow a .062" drill to be inserted between valve and body as shown in the illustration.

# SERVICE HANDBOOK

### SILVER WRAITH — SILVER DAWN — BENTLEY MK. VI.
### R. TYPE BENTLEY — PHANTOM IV.

Having set the kick diaphragm and the air gap, fix the housing in a vice by means of two ½" bolts and nuts fitted to the flange holes, connect a 12 volt battery to the solenoid and adjust. Adjust the spring tension so that the weight opens the choke to the amount required. A special tool is designed for this purpose.

(4) Fast Idle Setting.

With the engine warm set the mixture and slow running in the usual way. Adjust the fast idle screw on the highest step of the cam to give .035 clearance between the throttle stop screw and the stop.

The Bi-Metallic Solenoid Switch.

A further modification has been incorporated by the introduction of a Bi-Metal switch situated on the conduit elbow on the R.H. side of the dashboard just below the fuse box and wired in the solenoid circuit.

As the positive closing of the choke is unnecessary when the under bonnet temperature is more than 15° C (approx) this switch "breaks" at this temperature and recloses at 10° C. The system therefore ensures full closing for cold starting and avoids over-richness when the engine is hot or warm.

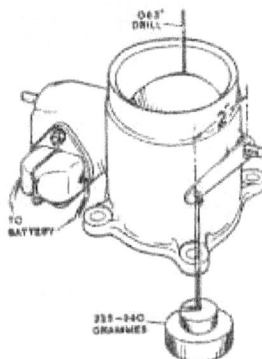

SETTING THE SPRING TENSION.

SERVICE HANDBOOK

SILVER WRAITH — SILVER DAWN — BENTLEY MK. VI.
R. TYPE BENTLEY — PHANTOM IV.

# SECTION
# D
# COOLING SYSTEM

# SERVICE HANDBOOK

### SILVER WRAITH — SILVER DAWN — BENTLEY MK. VI
### R. TYPE BENTLEY — PHANTOM IV.

## SECTION  D.

## COOLING SYSTEM

COOLANT - ENGINE THERMOSTAT - COOLANT PUMP SERVICING

- RADIATOR FAN - FAN BELT ADJUSTMENT - OVERHEATING

- RADIATOR - RADIATOR CALORSTAT - RADIATOR SHUTTERS

- RADIATOR STEAM VALVE - CAR HEATERS.

# SERVICE HANDBOOK

SILVER WRAITH — SILVER DAWN — BENTLEY MK. VI.

R. TYPE BENTLEY — PHANTOM IV.

## SECTION D.

## COOLING SYSTEM

### SILVER WRAITH — SILVER DAWN — BENTLEY MK. VI.
### R. TYPE BENTLEY — PHANTOM IV.

#### COOLING SYSTEM

| | |
|---|---|
| Capacity | - 4 gallons, (Silver Wraith, Silver Dawn, Bentley) |
| | - 5½ gallons, (Phantom IV) |
| Type | - Pressure |
| Pump | - Centrifugal |
| Pump drive | - "Vee" belt, 45.62" x .920" |
| Fan | - 5 blade |
| Radiator | - Thermostatically controlled shutters and by-pass thermostat (Silver Wraith, Phantom IV). Fixed shutters and by-pass thermostat (Silver Dawn and Bentley ) |

#### COOLANT:

The cooling system is filled with a 25% mixture of inhibited ethylene glycol and water on leaving the factory. This, or a similar solution of "Bluecol" should be used all the year round, the amount of anti-freeze being adjusted to suit more severe winter conditions as under.

#### INHIBITED ETHYLENE GLYCOL OR BLUECOL

| PERCENTAGE CONCENTRATION | 25% | 30% | 35% | 40% |
|---|---|---|---|---|
| Freezing point Fahrenheit | $10^{\circ}$ | $4^{\circ}$ | $-3^{\circ}$ | $-9^{\circ}$ |
| Degrees of frost Fahrenheit | $22^{\circ}$ | $28^{\circ}$ | $35^{\circ}$ | $41^{\circ}$ |
| Degrees of frost Centigrade | $12^{\circ}$ | $16^{\circ}$ | $19^{\circ}$ | $23^{\circ}$ |
| Silver Wraith } Silver Dawn } Bentley } | 8 pts. | 9½ pts. | 11 pts. | 12½ pts. |
| Phantom IV | 11 pts. | 13 pts. | 15½ pts. | 17½ pts. |

A similar solution must be used for topping-up.

If a suitable hydrometer is available it is possible to check the percentage of anti-freeze in the coolant, so that topping-up may be done with more anti-freeze or water to maintain the correct ratio required.

To check, note temperature of coolant with thermometer, take specific gravity reading with hydrometer, refer to Fig. D1, and read off degree of concentration.

Example:- A specific gravity reading of 1.05 at a temperature of $35^{\circ}$C indicates that 40% of the coolant is anti-freeze.

#### ENGINE THERMOSTAT:

The purpose of the thermostat is to ensure rapid warming-up of the engine and the maintainance of the correct coolant temperature.

SILVER WRAITH — SILVER DAWN — BENTLEY MK. VI.

R. TYPE BENTLEY — PHANTOM IV.

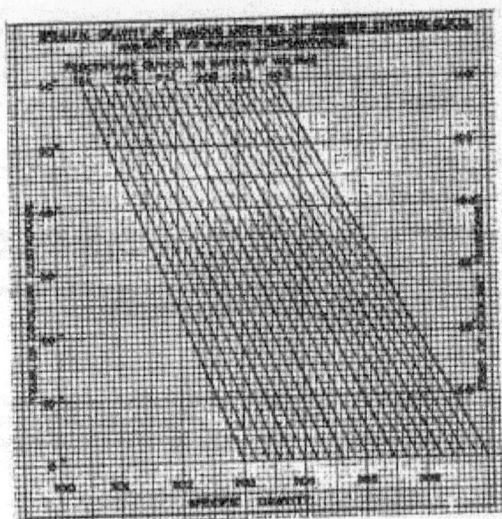

FIG. D1. COOLANT - SPECIFIC GRAVITY.

Fig. D2, shows thermostat exploded from housing.

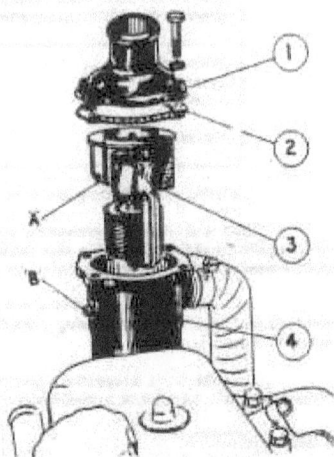

1. Outlet cover

2. Joint

3. Thermostat unit

4. Casing

A. Locating slot

B. Locating screw
   (Air release hole)

FIG. D2. THERMOSTAT.

The thermostat valve should start to open at $79° \pm 1°C$ and should be fully open at $95°C$.

## SILVER WRAITH — SILVER DAWN — BENTLEY MK. VI.
## R. TYPE BENTLEY — PHANTOM IV.

If a faulty thermostat is suspected, remove from housing and place in a suitable container with water and an accurate thermometer. On heating the water note the temperature at which the valve begins to open. Renew thermostat if not functioning correctly.

> NOTE:- The thermostat may be found to be stuck in its housing. Do not try to get a screw-driver under the valve and prise it out, but screw two 3 BA setscrews into the tapped holes provided on the top of the unit body and prise unit from housing with screw-driver under setscrew heads.

When replacing thermostat unit ensure that the slot 'A' in the body locates with the screw 'B' on the side of the thermostat housing. See Fig. D2.

When refilling the system after complete drainage, it is as well to temporarily remove the locating screw 'B', this will act as an air release hole; replace screw as soon as coolant spills freely.

### COOLANT PUMP:

The coolant pump is of the centrifugal type, and is combined with the radiator fan to form one belt-driven unit mounted on a common spindle. The spindle and the double-row ball bearing are integral, the bearing being packed with a special high-melting point grease at the time of manufacture, and requires no further lubrication.

Pressed on the front end of the spindle is the fan and pulley unit, whilst the pump impeller is pressed on the rear end, an inter-ference fit between the spindle and bore being the only means of re-tention.

The seal against seepage of coolant along the spindle consists of a packless gland assembled between the front end of the impeller and the casing. In the event of leakage it should be replaced.

1. Fan
4. Fan retaining screws
5. Spring ring
6. Grease retainer
7. Bearing and spindle
10. Pulley
11. Belt
14. Spring ring
15. Gland ring
16. Seal
17. Housing, seal
20. Spring
21. Flinger
23. Rotor

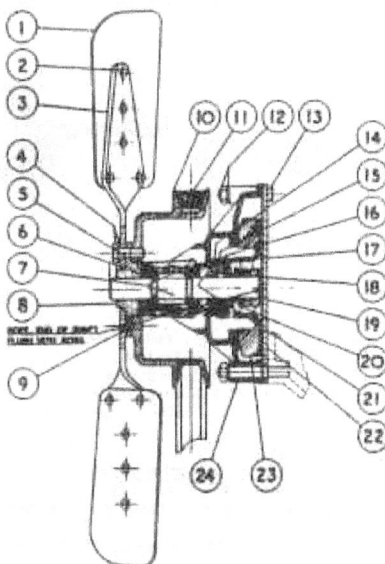

FIG. D3. COOLANT PUMP.

SILVER WRAITH — SILVER DAWN — BENTLEY MK. VI.

R. TYPE BENTLEY — PHANTOM IV.

COOLANT PUMP SERVICING:

To dismantle coolant pump - remove backing plate and with the aid of extractor RF-6025, withdraw the rotor from the spindle. Fig. D4.

Remove the circlip from the rotor, and then remove the gland ring and spring etc.

Examine the face of the carbon gland ring for wear and its projections for chatter marks, a new gland ring should be fitted if this is not in good condition.

Examine the carbon gland bearing facing on the pump casing, this should be refaced if scored. Squeaks and groans from the coolant pump usually emanate from wear on these two surfaces.

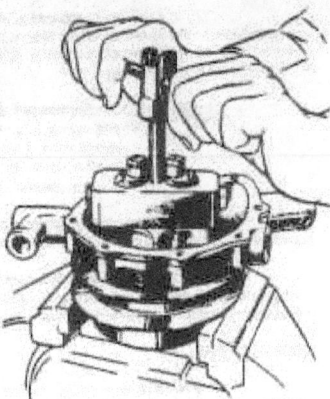

FIG. D4.  REMOVING ROTOR FROM SPINDLE.

Thickness of carbon gland ring    - .185" to .177" Max.
Width of carbon gland ring splines - .432" to .428" Max.

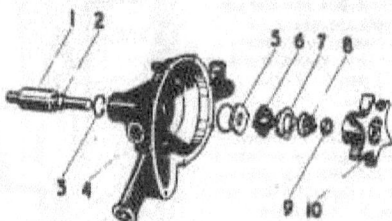

FIG. D5.  COOLANT PUMP.

1. Spindle and bearing.      6. Seal, rubber.
2. Flinger.                  7. Housing, seal.
3. Spring ring.              8. Spring.
4. Bearing face.             9. Retaining ring, seal.
5. Gland ring, carbon.      10. Rotor.

To fit new pump spindle - with the aid of the extractor RF-6025 withdraw the pulley and adaptor from the spindle. Remove the retaining spring ring from the casing and then tap out the spindle and bearing assembly.

Remove the flinger, light interference fit, and re-use on new spindle.

Remove the adaptor from the fan pulley, this and the spindle bearing assembly should be discarded as scrap. The old adaptor must not be used on the new spindle as some inevitable loss of interference may result in the fan and pulley working loose at a later date.

- D.4 -

Press flinger on to longer end of new spindle, .300" ± .015" from the outer end of the bearing.

Lubricate outer race of bearing and press spindle and flinger assembly into pump casing. Check flinger is not fouling. Fit retaining spring ring.

Fit new adaptor to fan pulley and secure.

Press the pulley and adaptor assembly onto spindle. Press on until .125" ± .005" of the spindle protrudes through the boss of the adaptor.

Using new gland ring and seal, complete the re-assembly.

FIG. D6.  REMOVING PULLEY FROM SPINDLE.

RADIATOR FAN:

As previously mentioned, the fan is attached to a pulley on the coolant pump spindle, and is driven by the same belt which drives the dynamo.

If the engine has to be run with the fan removed, it is essential that the fan retaining setscrews and washers with distance pieces .016" thick, to allow for the thickness of the fan, are fitted to retain the pulley.

The original fan fitted to the Silver Wraith, Silver Dawn and Bentley MK. VI engine was 16.00" dia., and revolved at 1.1 times engine speed. This has now been superseded by a larger diameter fan 17.750" running at .850" times engine speed. These two fans are not interchangeable.

The fan fitted to the Phantom IV, is 16.00" dia., and runs at 1.1 times engine speed.

FAN BELT ADJUSTMENT:

This is effected by releasing the dynamo retaining nuts and moving the dynamo outwards or inwards to suit.

If the belt is too loose, it will slip and wear excessively - if too tight it will cause premature wear to the pump spindle and dynamo bearings.

The tension should be such, that the fan belt can be moved 1" transversely, with finger, (i.e. half an inch either way) when checked at a point equidistant from the crankshaft pulley and the fan pulley.

To change a belt - do not strain over pulleys, slack off nuts and move dynamo fully inwards.

OVERHEATING:

Overheating may be due to one or more of the following causes:-

(a) Radiator calorstat may have failed (if fitted)
(b) Shutter mechanism may have jammed (if fitted)

SILVER WRAITH — SILVER DAWN — BENTLEY MK. VI.

R. TYPE BENTLEY — PHANTOM IV.

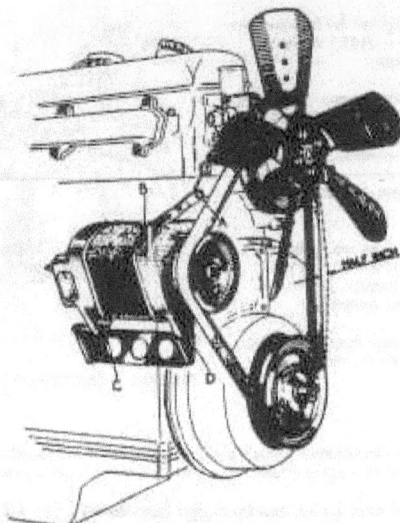

FIG. D7. ADJUSTMENT OF FAN BELT.

(c) Fan belt may need adjustment.
(d) Shortage of coolant in system.
(e) Detonation and poor grade of fuel.
(f) Radiator matrix blocked either externally or internally.

RADIATOR:

The radiator comprises two units, the outer shell, carrying the radiator shutters, and the matrix itself, the complete assembly mounted on a central rubber support.

The shell, diagonally braced, is connected to the wings and valances, with the matrix secured in the shell at three points.

Early models were fitted with the Mk.VI type matrix, later superseded by the Mk.VII type as below:-

| | | |
|---|---|---|
| Silver Wraith | - WTA-1 to WVH-116<br>ALW-1 to HLW- 41 } | Mk.VI type matrix. |
| | HLW-43 and onwards | Mk.VII type matrix. |
| Silver Dawn | - All series, | Mk.VII type matrix. |
| Bentley | - B-2-AK to B-270-DA, | Mk.VI type matrix. |
| | B-272-DA and onwards, | Mk.VII type matrix. |
| Phantom IV | - All series, | Mk.VII type matrix. |

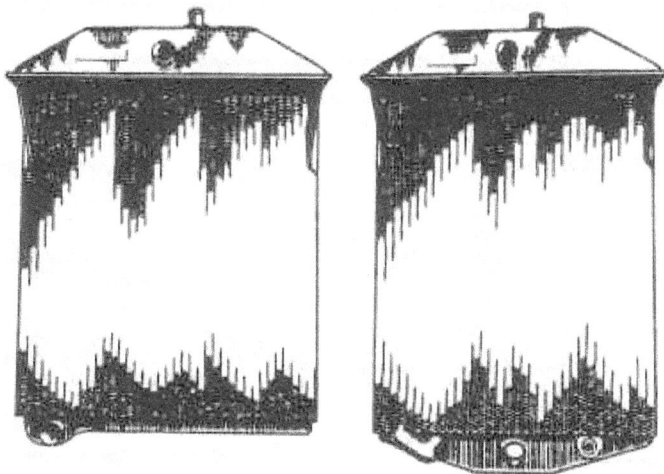

RADIATOR RECOGNITION.

FIG. D8.  MK.VI TYPE MATRIX.          FIG. D9.  MK.VII TYPE MATRIX.

## RADIATOR CALORSTAT:

Fitted to the Silver Wraith and Phantom IV models, to operate the shutters in front of the radiator.

The calorstat stem should start to move when the temperature of the coolant in the radiator header tank reaches 75°C and give the full .425" travel at 90°C. Also, from the inner stop when a temperature of 90°C is reached, the stem should move a load of 27½ lbs. + or – 2.

Reference to the instrument board thermometer will indicate if the calorstat is operating correctly.

## RADIATOR SHUTTERS:

Movable shutters are fitted to the Silver Wraith and Phantom IV models whilst fixed shutters are fitted to Silver Dawn and Bentley.

In the event of faulty operation of the radiator calorstat with consequent failure to open the shutters, disconnect the calorstat by raising the spring loaded pin (2) and disengaging the end of the lever (3) from the calorstat rod. A knob (4) is provided to ease re-engagement.

FIG. D11.  RADIATOR SHUTTER CALORSTAT.

The shutters must be pushed open by means of the lever (3) and not by direct hand operation.

If the pin (2) should have been disengaged when the calorstat is in working order it will be difficult to re-engage with the system cold. If the engine is first warmed up to a temperature at which the shutters would normally commence to open, re-engagement will be found to be easy.

## RADIATOR STEAM VALVE:

Originally, all Silver Wraith and Bentley Mk.VI radiators incorporated a steam valve in the overflow pipe from the top tank; this put a pressure of 4 lbs/sq.in. on the system. Later, a modification was introduced deleting the valve and spring, and operating the coolant system at atmospheric pressure.

Later, owing to complaints of loss of coolant, a reduced pressure valve was introduced and is now fitted to all models. See Fig. D12.

Spring particulars:-

    Free length - .900"
    Compressed to .600" - 3 ozs.

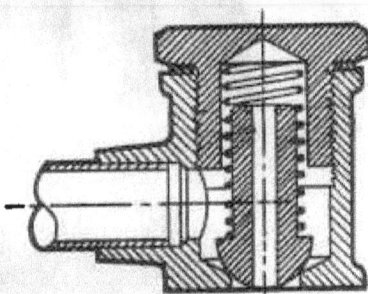

FIG. D12. RADIATOR STEAM VALVE.

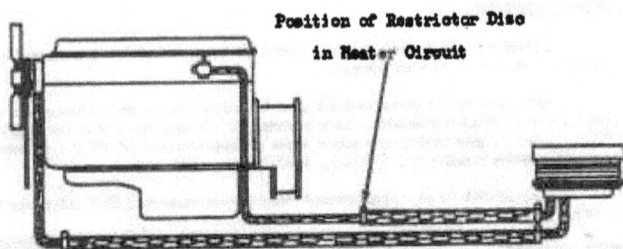

Position of Restrictor Disc
in Heater Circuit

FIG. D13. HEATER CIRCUIT.

## CAR HEATERS:

An electrically operated hot water heater is fitted as standard equipment on all models under the front passenger's seat.

A tap, fitted into the cylinder head, supplies the hot coolant for the heater matrix, and this should be used to isolate the system when not required.

- D.8 -

### SILVER WRAITH — SILVER DAWN — BENTLEY MK. VI.
### R. TYPE BENTLEY — PHANTOM IV.

To obviate building up too high a pressure in the system, a restrictor is fitted, see Fig. D13. When customer requirements call for the fitting of an extra heater under the dash, the position of the restrictor should be removed as shown in Fig. D14.

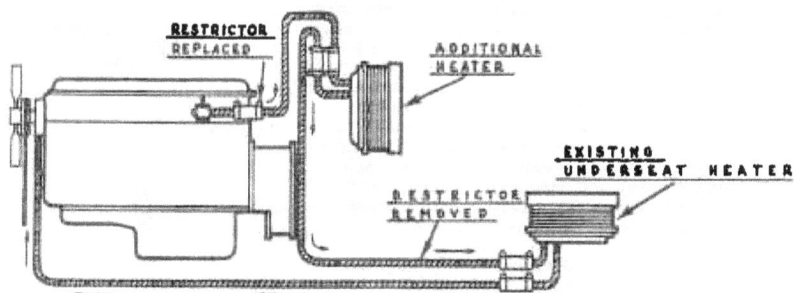

FIG. D14. HEATER CIRCUIT - UNDERDASH HEATER.

SILVER WRAITH — SILVER DAWN — BENTLEY MK. VI.

R. TYPE BENTLEY — PHANTOM IV.

## RADIATOR FLOW TESTING.

The radiator may be checked by means of a water flow test. A container with a 1" bore pipe let into the bottom should be hung above the radiator and connected to the header tank intake pipe, the radiator lower connection being allowed to run to waste. The tank should be suspended so that the centre is 3-ft above the radiator intake pipe, when the time taken for the tank to empty will give an approximation of the condition of the matrix. The figures given below do not include the time required for the radiator to empty after the flow from the tank has ceased. The water level in the tank may be observed if the top is cut open; a mirror attached at a suitable angle would permit observation from ground level, or by means of a glass observation tube fitted as shown in the illustration and secured with jubilee clips.

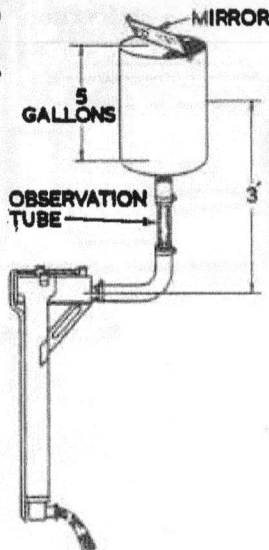

FLOW TEST RIG.

| Container. | Mk.VI Rad. | Mk.VIII Rad. |
|---|---|---|
| 5 Gall. (Oil Drum) | 14 secs. | 22 secs. |
| 10 Gall. (Tank) | 28 " | 44 " |
| 18 Gall. (Petrol Tank) | 50 " | 80 " |

These flow times are for a radiator in new condition.

NOTE:    The Mk.VII radiator which was fitted
(BENTLEY ONLY)  on Chassis No. B-270-DA and onwards, incorporates a drain tap and a starting handle tunnel in the bottom tank, neither of which is featured on the earlier type.

Should the flow time exceed those quoted by over 25% it is probable that the radiator is the cause of overheating, and should be replaced.

When fitting a replacement radiator, it is essential to flush out the engine thoroughly. The thermometer and thermostat should be carefully checked, as these may be damaged by severe overheating.

A radiator with a flow figure of less than 25% above the nominal should be given a Cooling System Service as a preventive measure, but it is not likely to be the cause of overheating in this condition.

## Annual Routine Servicing.

Drain the system and flush out as follows :-

1. Remove the top and bottom radiator hoses and reverse flush the radiator with mains pressure for ½ hour.

2. Remove the cylinder block drain tap and thermostat or thermostat housing and flush from both ends for about ¼ hour each.

3.   Refill the system.

3a.  Smith's "Bluecol" anti-freeze and the Ethylene Glycol
     used at the Factory now both contain a new inhibitor,
     but before using either in a car which has been filled
     with the earlier anti-freeze, the following action is
     necessary to conserve the new inhibitor <u>which will
     otherwise be consumed within a few hours</u>.

   a)   Start the engine and run for a few minutes.

   b)   Remove cylinder drain cock and insert a hose into
        the radiator, run the engine, occasionally opening
        the throttle slightly until the water runs clear;
        this usually takes about 20 minutes.

   c)   Replace the drain cock, add ½ pint of Stergene to
        the full system and <u>run for 5 minutes at normal
        engine temperature</u>.

   d)   Repeat (b) and run until cool, then switch off and
        leave until engine is cold.  Start up and run until
        the water ceases frothing.

   e)   Drain and refill with anti-freeze.


           This operation need only be done once on any car, provided
that the owner continues to use the recommended anti-freeze.  A distinct
fluorescent effect can be observed in the new type proprietary anti-freeze; the
Ethylene Glycol equivalent is yellowish-green.

SILVER WRAITH — SILVER DAWN — BENTLEY MK. VI.

R. TYPE BENTLEY — PHANTOM IV.

## THE HEATER AND DEMISTING SYSTEMS.

As a result of investigation and development the Demister system has been modified from time to time. The various schemes employed are given hereunder. These will be helpful in identifying the types that may be expected with certain chassis numbers. But, it should be noted that in some cases the earlier types have been removed and replaced with later types.

(i)     "Bentley Mk.VI B2AK - B191LZ
        Silver Wraith  WTA1 - WDC99
        Not fitted to LH drive cars."

Demister motor and air intake situated under scuttle. Defroster heater element in RH side of outlet duct. Controlled by switches on facia. Switch marked "M" to operate motor and switch marked "F" to operate Defroster.

(ii)    "Bentley Mk.VI B193DZ - B271EW
        Silver Wraith  WDC100 - WGC49."

Demister motor situated under scuttle. Radiator heated air intake with tubing connections between air intake and Demister motor. Controlled by means of "Eyelids" on windscreen capping rail, with the switch "M" operating the motor. In this scheme the heater element is deleted.

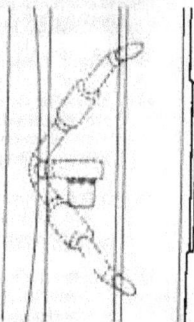

ORIGINAL SYSTEM.

(iii)   "Bentley Mk.VI B273EW - B149HP
        Silver Dawn    LSBA2 - LSCA51
        Silver Wraith  WGC50 - WLE25."

Radiator heated air intake, Demister motor on engine side of dash. Duct connecting air intake to Demister motor. Controlled by "Eyelids" and switch "M" to operate the motor.

(iv)    "Bentley Mk.VI B151HP - B251JN
        Silver Wraith  WLE26 - WSG101
        Silver Dawn    LSCA53 - LSDB4."

Radiator heated air intake. High speed Demister motor and modified transfer ducting to give improved air flow. The Demister motor bracket is deleted and the motor is suspended in the transfer ducting. The air intake is fitted with a deflector designed for use with the later type triple pass system (Scheme vii). Controlled by "Eyelids" and switch "M" for the Demister motor. This is a intermediate scheme prior to the introduction of the triple pass system.

# SERVICE HANDBOOK

### SILVER WRAITH — SILVER DAWN — BENTLEY MK. VI.
### R. TYPE BENTLEY — PHANTOM IV.

(v)     *Bentley Mk.VI B2KM - B51PU
        Silver Dawn    LSDB6- SHD60."

        Radiator heated air intake designed to pass air three times
through the radiator matrix thereby ensuring a higher air temperature.  Ducting
and Demisting motor as in scheme iv.

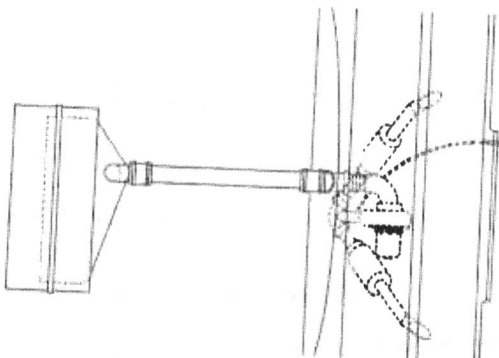

RADIATOR HEATED SYSTEM (INTERMEDIATE TYPE).

(vi)    "Bentley       B53PU - B401TO
        Silver Wraith WVH1  ~ WVH98  (Long wheelbase ALW27 - BLW42)
        Silver Dawn   SKE2  - SMF76."

        Similar to scheme v, but with hot and cold demisting
controlled by an under bonnet hand lever operating two spring loaded butterfly
valves, one of which opens when the other closes.  The cold air being drawn
through the RH dummy horn grill.

(vii)   "Bentley       B2TN - B244YD
        Silver Wraith BLW43- DLW79   (Long wheelbase)
        Silver Dawn   SNF1 - SUJ12."

        Similar to scheme v but with a wire cable control from the
instrument board operating the valve lever.

(viii)  *Bentley       B246YD - onwards
        Silver Wraith DLW80  ~ onwards
        Silver Dawn   SUJ14  - onwards

Continental Bentley Only.

        A flexible rubber and wire duct from a grill in the front
wing to a manually operated trap to the drivers compartment provides fresh air
ventilation.

SILVER WRAITH — SILVER DAWN — BENTLEY MK. VI.

R. TYPE BENTLEY — PHANTOM IV.

For the windscreen, a separate air tube is tapped into this ducting and taken through a heater fed from the engine cooling system, to the motor mounted on the front of the dashboard, then through ducts to vents in the capping rail (These vents are fitted with "Eyelids").

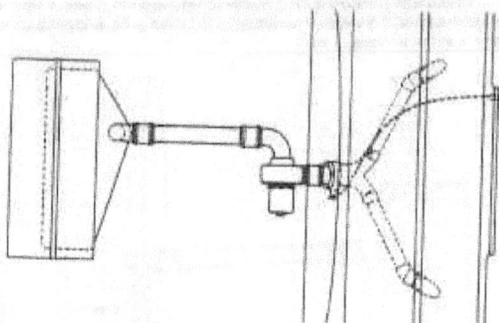

IMPROVED TYPE DEMISTER SYSTEM.

## The Rear Window Demister.

The rear window.demister was introduced with the following chassis numbers :

Bentley B311NY
Silver Dawn SFC94

and consists of an electrically heated glass pane, with a series of wires moulded into the glass. It is controlled from the facia board by the switch "RW".

## Test Procedure.

Create artificial misting by means of an electric kettle, with all windows closed, keeping the exterior of the windscreen cool with an air blast or sprinkler hose. Warm the engine to normal running temperature, remove the kettle and operate demister.

If the ducts are free the screen will clear except for a narrow "V" which will clear gradually.

If the vents or ducts are obstructed, the "V" pattern will be broad, as shown.

UNSATISFACTORY PATTERN

The source of obstruction can then be traced by starting from the "Eyelids" and working downwards.

- D.14 -

Servicing.

In complaints of inefficiency of the demisting system, the following items should be checked:-

Demister Motor.

Clean brushes, check the spring tension and also free movement. Clean commutator with petrol moistened cloth.

Air Leaks.

Check for leaking at joints or piping out of alignment.

Air Flow Restriction.

SATISFACTORY PATTERN

The slot in the capping rail must be in alignment with the Duct delivery aperture, and give an unobstructed passage to the air flow. Dirt frequently obstructs the flow.

"Eyelids."

The "Eyelids" must open sufficiently, check that they are not bent. It may be found necessary to cut the woodwork beneath the capping rail to correct these faults.

Delivery Pattern on Windscreen.

An unsatisfactory cleared pattern on the windscreen is usually due to one of the above, but quite often it is caused by incorrect shaping of the duct at its delivery point.

The ideal shape is a 1/16 slot, with the duct main body left as full as possible, especially round the windscreen wiper.

MAINTAIN $\frac{1}{16}$" GAP CLEAR

CAPPING RAIL DUCT AND EYELIDS.

SERVICE HANDBOOK

SILVER WRAITH — SILVER DAWN — BENTLEY MK. VI.
R. TYPE BENTLEY — PHANTOM IV.

# SECTION
# E
# CLUTCH

# SERVICE HANDBOOK

### SILVER WRAITH — SILVER DAWN — BENTLEY MK. VI.
### R. TYPE BENTLEY — PHANTOM IV.

## SECTION E.

## C L U T C H

### List of Illustrations:-

## C L U T C H

| | | |
|---|---|---|
| Make | - | Borg and Beck |
| Type | - | Single dry plate |
| Size | | |
| Silver Wraith | - 11" | (Heavy) |
| Silver Dawn | - 10" | (Long) SBA.2 - SCA.25 |
| | 11" | (Light) SCA.27- SDB.74 |
| | 11" | (Heavy) SDB.76- |
| Bentley | - 10" | (Long) B.2.AK- B.401.GT |
| | 11" | (Light) B.2.HR- B.298.LJ |
| | 11" | (Heavy) B.300.LJ - |
| Phantom IV | - 11" | (Heavy) |
| Facing material | - | Mintex H.14 |
| Pedal free travel | - 10" | (Long) $1\frac{1}{4}$" - $1\frac{1}{2}$" |
| | 11" | (Light) $\frac{3}{4}$" - 1" (short lever) |
| | 11" | (Heavy) $1\frac{1}{4}$" - $1\frac{1}{2}$" |

As an easy means of recognition between the 11" "Light" and 11" "Heavy" type of clutch, the clutch external operating lever for the "Heavy" type is 4.250" long, between centres, whereas for the "Light" type it is 3.625". Thus the difference in clutch pedal adjustment, see above.

CLUTCH PEDAL ADJUSTMENT:

As the driven plate facings wear, the pressure plate moves closer to the flywheel and the weighted ends of the three release levers follow. This causes the inner ends of the release levers to travel further towards the gearbox and decrease the clearance between the levers and the clutch release bearing. The effect is to decrease the free travel of the clutch pedal.

Periodically check and adjust to correct dimensions, see above.

CLUTCH SPRINGS:

Number of pressure springs - 9.

The original 10" clutch was fitted with "Orange" coloured pressure springs. These have now been superseded by "Red" coloured springs, and whenever these units are being overhauled a check should be made that "Red" springs are fitted.

"Red" springs - Load required to compress to 1 9/16" = 150 - 155 lbs.

The 11" Light type and the 11" Heavy type are both fitted with "Orange" coloured pressure springs, except for Phantom IV.

"Orange" springs - Load required to compress to 1 9/16" = 130-140 lbs.

The 11" Heavy type when fitted to Phantom IV is fitted with "Red" coloured pressure springs as above specification.

The driven plate damper springs are "Red" coloured for all series except Phantom IV, which should be "Yellow".

**SILVER WRAITH — SILVER DAWN — BENTLEY MK. VI.**

**R. TYPE BENTLEY — PHANTOM IV.**

Damper springs:-

"Red" - Load required to compress to 15/16"  = 140 - 154 lbs.

"Yellow" - Load required to compress to 15/16"  = 155 - 169 lbs.

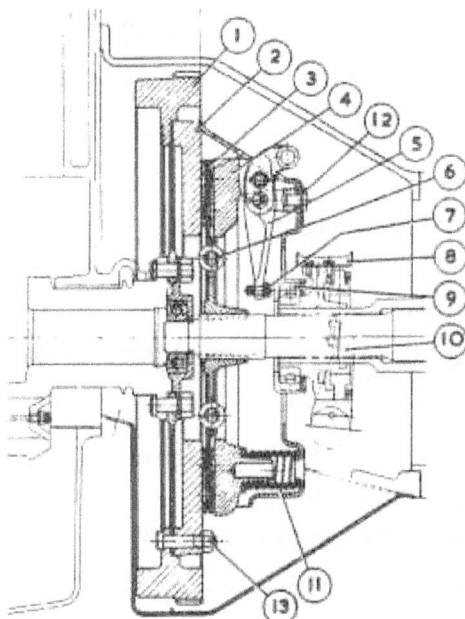

FIG. E1. SECTION - CLUTCH.

1. Flywheel
2. Friction plate
3. Driven plate
4. Pressure plate
5. Release lever
6. Damper spring
7. Adjuster screw
8. Release bearing
9. Oil trough
10. Trunnion
11. Pressure spring
12. Setscrews
13. Locked bolt

SILVER WRAITH — SILVER DAWN — BENTLEY MK. VI.

R. TYPE BENTLEY — PHANTOM IV.

RELEASE BEARINGS:

Ball Bearing - 1.75" x 3.00" x .562".

Lubrication of ball bearing and trunnion, from chassis Luvax system.

In the event of the existing clutch release bearing becoming noisy it should be renewed. Access to the bearing is obtained by the removal of the gearbox, but it is <u>not</u> necessary to disturb the clutch.

SERVICE FAULTS:

| Cause | Remedy |
|---|---|
| **Clutch Slip.** | |
| Insufficient free travel | - Adjust clutch pedal |
| Weak thrust springs | - Fit replacement cover unit |
| Worn driven plate | - Renew |
| Scored pressure plate | - Reface - see page E.6 |
| Oil on driven plate | - Renew, check crankshaft oil seal |
| **Clutch Shudder.** | |
| Burnt oil on facing | - Renew driven plate |
| Incorrect adjustment of release levers | - Check and adjust |
| Weak driving springs | - Renew driven plate |
| Worn driven plate | - Renew driven plate |
| **Clutch Spin or Drag.** | |
| Too much free travel | - Adjust clutch pedal |
| Insufficient total travel | - Adjust clutch pedal |
| Incorrect adjustment of release levers | - Check and adjust |
| Distorted driven plate | - Renew driven plate |
| **Clutch Rattle.** | |
| Driving springs broken or weak | - Renew driven plate |

CLUTCH REMOVAL:

The recommended procedure is to first remove the gearbox:-

(1)    Disconnect speedometer drive and flexible oil pipe from rear of gearbox.

(2)    Disconnect oil damper control rod.

(3)    Slacken back inner nuts (4, Fig. E2), and remove torque reaction rubbers then remove torque bracket.

(4)    Remove tie-bar complete with bracket and packing piece.

(5)    Disconnect centre universal joint; disconnect front propellor shaft from gearbox, slide clear.

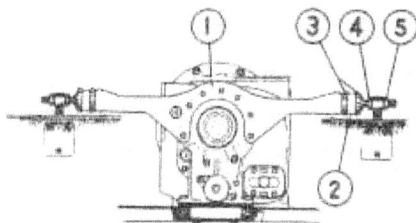

FIG. E2. GEARBOX REAR MOUNTING.

1. Gearbox torque bracket.
2. Torque reaction rubbers.
3. Support cups.
4. Inner nuts.
5. Outer nuts.

(6)   Uncouple change gear lever, remove selector shaft from box.

(7)   Remove pull rods and drag links from servo, ease gearbox sideways and remove servo motor from driving shaft.

(8)   Support rear of engine with jack and draw away gearbox.

If difficulty is experienced in withdrawing first motion shaft from spigot bearing, see appropriate chapter in Gearbox Section.

It is not possible to remove clutch casing from engine whilst in the frame, owing to dashboard preventing it being lifted over flywheel, therefore, extract clutch from casing.

(1)   Remove bottom cover and turn flywheel until one of the three locked bolts (13, Fig. E1), retaining friction plate to flywheel, is at lowest position. Remove nut and bolt to provide required withdrawal clearance.

(2)   Turn flywheel to expose two of the six nuts retaining clutch to flywheel. Slacken off, and at the same time insert between release lever and cover, a Borg and Beck "L" shaped spacer, if not available, use a ¼" B.S.F. nut. Remove bolts.
Turn flywheel and repeat above operations to remove remaining nuts.

(3)   Hold clutch, and turn flywheel to bring bolt hole (13) to lowest position and remove clutch, complete with driven plate.

## CLUTCH COVER ASSEMBLY:

The assembly need not be dismantled for inspection. The pressure plate should be free from deep scoring. If scoring, distortion or surface cracks are evident, reface pressure plate or fit a new assembly.

The release lever assembly shows a certain amount of slackness even when new, if necessary, the thrust (pressure) springs should be checked as per specification on page E1.

SILVER WRAITH — SILVER DAWN — BENTLEY MK. VI.

R. TYPE BENTLEY — PHANTOM IV.

To dismantle - Mark the following parts to ensure identical re-assembly and balance, (a) cover, (b) pressure plate, (c) release levers.

(1) Place cover assembly on the bed of a press with pressure plate on blocks to allow free movement of cover when depressed.

(2) Place wooden block across cover, resting on spring bosses, and operate press. Remove the setscrews (12, Fig. E1) and then release pressure slowly, remove cover and collect thrust springs and washers.

(3) Extract split pins, push out pins and collect yokes and rollers. Remove release levers and collect needle bearings.

To re-assemble - Reverse the above instructions, ensuring that all co-relation marks coincide.

Before fitting new clutch driven plate, check plate for parallelism of faces. A simple jig can be made as illustrated in Fig. E3.

Limit of out of parallel - .012".

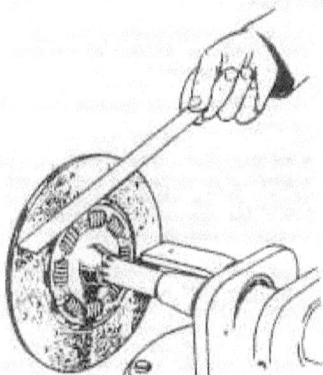

FIG. E3.   JIG FOR CHECKING PARALLELISM OF FACES.

External diameter of plate  -  8.400"
Thickness of machined lugs  -  .327" to .330"
Height of central hub       -  2.062"
N.B. A special extension piece may be
   necessary to raise existing hub
   to this dimension.

Release Lever Adjustments:

Clearance, 2.062" ± .030", between flywheel face and top of hemispherical headed adjusting screws.

Maximum variation between screws  -  .005".

For accurate adjustment, a Borg and Beck gauge plate is essential. Fig. E4.

The gauge plate should conform to the following dimensions:-

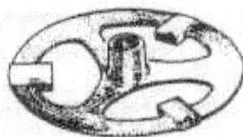

FIG. E4.   BORG & BECK LEVER ADJUSTMENT GAUGE.

(1) Position gauge plate centrally against pressure plate, with machined lugs below release levers, bolt complete assembly to surface plate using full compliment of retaining bolts. Check with thin feeler to ensure that cover is tightened down squarely.

# Service Handbook

## SILVER WRAITH — SILVER DAWN — BENTLEY MK. VI.
## R. TYPE BENTLEY — PHANTOM IV.

(2)  Place a short straight edge across top of central hub, and adjust each release lever separately, the top of the hemispherical headed screw should just make contact with the straight edge. Lock by peening metal into saw cuts.

(3)  Replace spacers for re-erection.

FLYWHEEL FRICTION AND CLUTCH PRESSURE PLATES:

Both these plates may be found scored or subject to slight contraction cracks. It is permissible to regrind to a depth not exceeding .010" on each plate from the original dimensions, as under:-

Flywheel friction plate  -  .800" - 10 (A. Fig. E5)
Clutch pressure plate
10" C.F.       - 1.284" - 6 )
11" Light )   - 1.286" - 10 )(B. Fig. E5)
11" Heavy )

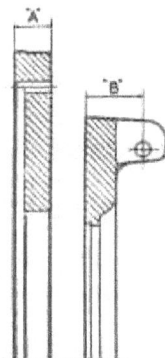

RE-ASSEMBLY:

(1)  Check the replacement driven plate for freedom on splines of gearbox first motion shaft.

(2)  Re-assemble in the reverse order of removing.

(3)  Make sure that the oil feed pipe is correctly positioned above the oil trough on the clutch trunnion, and check the rate of oil flow to the clutch release bearing.

Depress the foot pedal of the oil pump once, and count the number of drops delivered, three to five drops should be delivered at each application of the pedal. If the rate of flow is greater or less than this, re-new the restrictor elbow. Note the correct size number(they are lettered and numbered to indicate their shape and rate of flow, a higher number indicating a greater rate of flow)

FIG. E5.  FRICTION AND PRESSURE PLATES.

(4)  Refitting gearbox - adjusting torque reaction rubbers - replace rubbers in their cups, recessed ends outermost, and then tighten up inner nuts to lock, holding the cups square while doing so. DO NOT disturb the two outer nuts.

NOTE:  If the outer nuts have been disturbed or it is necessary to reset the adjustment:-

Fully slacken back the inner and outer nuts, and correctly replace the rubbers, making sure that these are fully pressed home in their cups. Tighten up the inner nuts finger tight, then holding cups square, tighten two more complete turns with spanner. Tighten outer nuts to lock.

After refitting flexible oil pipe to shock damper control, bleed the control system:-

Start engine and run slowly in top gear approximately 10 M.P.H., move Ride Control lever to "Hard" and then remove air release plug from damper.

Continue to run engine until a continuous flow of oil runs from damper. Replace plug.

Repeat for damper on opposite side.

Top up gearbox level.

# SECTION
# F
# GEARBOX

# SERVICE HANDBOOK

SILVER WRAITH — SILVER DAWN — BENTLEY MK. VI.
R. TYPE BENTLEY — PHANTOM IV.

SECTION F.

## GEARBOX

(Synchromesh)

GEARBOX RATIOS - GENERAL - SERVICE FAULTS - REMOVAL FROM
CHASSIS - DISMANTLING GEARBOX - RE-ASSEMBLING GEARBOX -
SIDE GEAR CHANGE CONTROLS - COLUMN GEAR CHANGE CONTROL.

## GEARBOX

(Automatic)

GEARBOX RATIOS - GENERAL - OPERATION - COASTING OR TOWING
- TOPPING-UP OR CHANGING FLUID - LINKAGE SETTINGS (EARLY
TYPE) - 1,000 MILE ADJUSTMENT.

SILVER WRAITH — SILVER DAWN — BENTLEY MK. VI.

R. TYPE BENTLEY — PHANTOM IV.

## SECTION F.

## G E A R B O X.

List of Illustrations:

List of Illustrations:     (Continued)

**SILVER WRAITH — SILVER DAWN — BENTLEY MK. VI.**

**R. TYPE BENTLEY — PHANTOM IV.**

SECTION F

GEARBOX

Oil Capacity - 6 pints
Weight      - 104 lbs.

Gearbox Ratio's:

|  | Standard | Close Coupled | Bentley Continental Sports Saloon |
|---|---|---|---|
| 1st Speed - | 2. 98 : 1 | 2.70 : 1 | 2. 67 : 1 |
| 2nd Speed - | 2.017 : 1 | 1.82 : 1 | 1. 54 : 1 |
| 3rd Speed - | 1. 34 : 1 | 1.22 : 1 | 1.216 : 1 |
| 4th Speed - | Direct | Direct | Direct |
| Reverse - | 3.156 : 1 | 2.86 : 1 | 2.861 : 1 |

Silver Wraith
Silver Dawn
Bentley

Overall Ratio's - with 11:41 Rear Axle (Standard).

|  | Standard | Close Coupled |
|---|---|---|
| 1st Speed - | 11.11 : 1 | 10.07 : 1 |
| 2nd Speed - | 7.52 : 1 | 6.81 : 1 |
| 3rd Speed - | 5. 0 : 1 | 4.53 : 1 |
| 4th Speed - | 3.73 : 1 | 3.73 : 1 |
| Reverse - | 11.76 : 1 | 10.66 : 1 |

Overall Ratio's - with 12:41 Rear Axle.

|  | Standard | Close Coupled |
|---|---|---|
| 1st Speed - | 10.19 : 1 | 9.23 : 1 |
| 2nd Speed - | 6.89 : 1 | 6.24 : 1 |
| 3rd Speed - | 4.58 : 1 | 4.15 : 1 |
| 4th Speed - | 3.42 : 1 | 3.42 : 1 |
| Reverse - | 10.78 : 1 | 9.77 : 1 |

Overall Ratio's - with 13:40 Rear Axle (Bentley Continental Sports)

| 1st Speed - | 8.222 : 1 |
|---|---|
| 2nd Speed - | 4.750 : 1 |
| 3rd Speed - | 3.741 : 1 |
| 4th Speed - | 3.077 : 1 |
| Reverse - | 3.802 : 1 |

Phantom IV

Overall Ratio's - with 8:34 Rear Axle (Standard)

| 1st Speed - | 12.74 : 1 |
|---|---|
| 2nd Speed - | 8.52 : 1 |
| 3rd Speed - | 5.71 : 1 |
| 4th Speed - | 4.25 : 1 |
| Reverse - | 13.4 : 1 |

# SERVICE HANDBOOK

SILVER WRAITH — SILVER DAWN — BENTLEY MK. VI.
R. TYPE BENTLEY — PHANTOM IV.

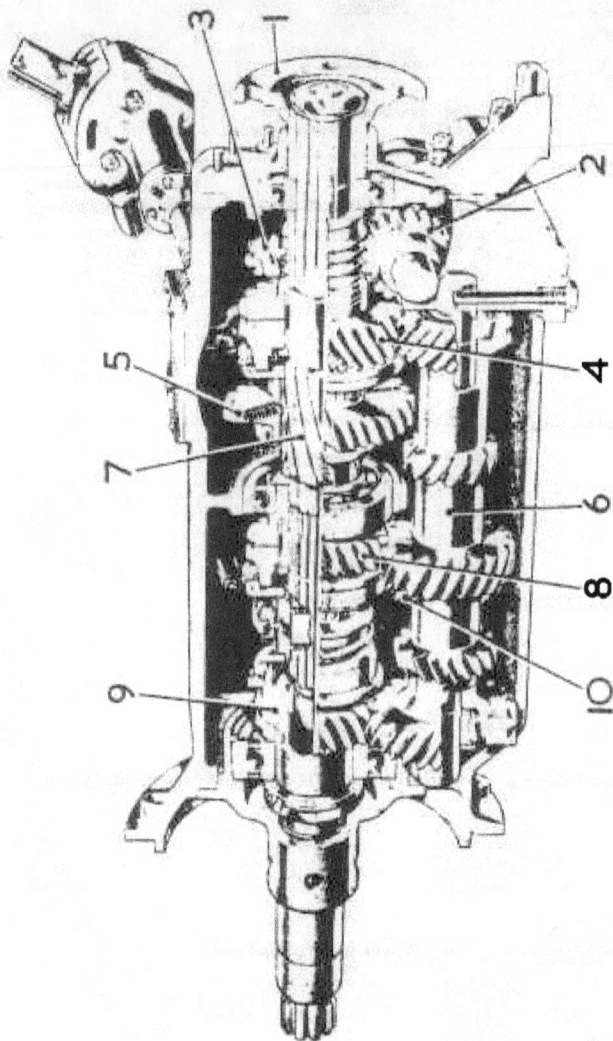

FIG. F.1. SECTIONAL VIEW OF GEARBOX.

1. Propeller Shaft Drive Flange.
2. Servo Drive.
3. Speedometer and Pump Drive.
4. 2nd Speed Constant Mesh Gear.
5. 1st Speed Gear.
6. 2nd Motion Shaft.
7. 3rd Motion Shaft.
8. 3rd Speed Constant Mesh Gear.
9. 1st Speed Constant Mesh Gear.
10. Reverse Gear.

SILVER WRAITH — SILVER DAWN — BENTLEY MK. VI.

R. TYPE BENTLEY — PHANTOM IV.

GENERAL:

The engine and gearbox are of unit construction, suspended at the front of the engine and rear of the gearbox. Torsional rigidity is controlled by the torque arm at the rear of the gearbox, bearing on two rubbers. The fore and aft location is obtained by a tie-bar.

The gearbox contains four forward speeds and reverse with synchromesh on second, third and fourth speeds. It is operated by either right-hand lever, centre lever, or column gear controls. Brake servo-motor and speedometer drives and a gear type pump for rear controllable shock dampers are mounted on the gearbox and driven from the third motion shaft.

SERVICE FAULTS:

Difficult Gear Change:

(i)     Check dryness or dirt in side gear mechanism. Remove gaiter and clean gate, etc.

(ii)    Incorrect adjustment of rear tie rod (Fig. F31) causing cross binding. Reset, see page F17.

(iii)   Synchromesh cones sticking, causing drag. Polish, see page F10.

(iv)    Stiffness or failure of sliding keys or grooved spheres of early type third motion shafts, difficult to change from third to top. Ease keys or fit new third motion shaft.

Jamming of Reverse Gear in Gearbox:

(i)     Incorrect setting of lever (9, Fig. F13). See page F8. To free bronze reverse actuating lever, slacken off the two nuts (beneath gearbox) securing locking plate of eccentric pivot pin, 7, Fig. F13, rotate pin half turn.

"Clonking" between Drive and Overdrive:

(i)     Excessive end float on first and third motion shafts, caused by spring retaining rings depleating under load.

        Fit square edged ball bearings to first and third motion shaft, see page F10.

(ii)    Excessive end float of second and third speed driven gears. Reduce end float or if necessary fit new bushes, see page F12.

Noisy First Speed Gears:

(i)     Damaged teeth, due to re-engaging clutch before teeth of gears are fully disengaged.

        Stone leading edge of teeth, or if badly damaged, fit new gear or gears.

        NOTE:-  If new second motion shaft is necessary, a new first motion shaft must be fitted to ensure silent running of constant mesh gears, as these are paired.

### Noisy Second Speed Gears:

(i)    In a case of bad whine from gears, a new second speed driven gear may be necessary.

        NOTE:- If a new second motion shaft cluster is fitted, a new first motion shaft will be necessary to ensure silence of mating gears.

(ii)    Uneveness of gear noise in over-run.
Result of wear of bushes, 5 and 21, see page F20.

### Noisy Third Speed Gears:

(i)    In a case of bad whine from gears, a new third speed driven gear may be necessary.

(ii)    Uneveness of gear noise on over-run.
Result of wear of bushes, 11 and 28, see page F20.

### Noisy Constant Mesh Gears:

A bad whine from the constant mesh gears will necessitate a new mated pair of first and second motion shaft.

## REMOVAL FROM CHASSIS:

(i)    See Section E - Removal of Clutch.
See Section G - Removal of Servo Motor.

(ii)    Difficulty may be experienced in removing gearbox due to first motion shaft being too close a fit in, a collapse of spigot bearing. Remove the front cover and withdraw box leaving first motion shaft in position.

A special extractor is available from the London Service Station, but should the hire of this be impracticable, or with the use of the extractor removal still prove not possible, it will be necessary to remove the crankshaft from the engine complete with the first motion shaft in position and remove this on a bench operation where more purchase can be obtained.

## DISMANTLING GEARBOX:

(i)    Remove plug and drain off oil.

(ii)    Attach gearbox on suitable stand.

(iii)    Remove top cover plate. If selector lever shaft is still in position, remove setscrews (R.H. side of box) securing bearing covers and withdraw shaft through side of gearbox.

(iv)    Remove dipstick.
Remove damper pump oil feed tube. Remove side cover plate.

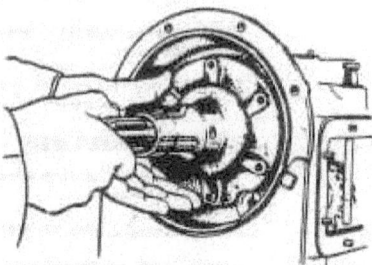

FIG.F2.   REMOVAL OR ASSEMBLY OF FIRST MOTION SHAFT.

(v)     Remove front end cover and withdraw first motion shaft assembly. Fig.F2.

NOTE:- There are 14 rollers and Bakelite roller retainer to recover. The retainer is used to assist assembly, and is pushed by the nose of the third motion shaft into hollow of first motion shaft. Remove the cone.

FIG. F4.    THIRD MOTION SHAFT IN POSITION - SHOWING RETAINING CIRCLIP.

FIG. F3.   REMOVING STAND PIPE.

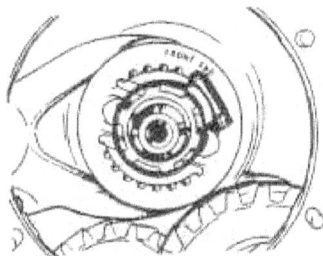

(vi)    Temporarily fit Jubilee clip to front end of third motion shaft against sliding piece, Fig. F4, to retain in position during removal of shaft.

(vii)   Remove nut, (1 Fig.F20) from rear end of third motion shaft. Remove lockwasher, two distance washers and coupling flange (3).

NOTE:- To prevent shaft turning while removing nut, insert two 7/16" bolts in coupling flange and hold with lever between bolts. See Fig. F14.

(viii)  Remove nuts securing rear end cover assembly. Remove torque bracket. Three tapped holes ($\frac{1}{4}$" x 26 TPI) are provided in cover, use three suitable set-screws 2$\frac{1}{4}$" long, fully threaded, screw into cover and remove progressively. Fig. F5.

NOTE:- Inner race of rear ball bearing (18 Fig. F20) is in two halves - front half remains on shaft. Now remove locating piece securing servo drive shaft. A 5/16" x 22 TPI hole is provided for extraction purposes. See Fig. F6. Next, remove countersunk headed screws (on RH side of gearbox) securing servo drive housing. Remove by using suitable drift on inner end of shaft and tap out. see Fig.F17.

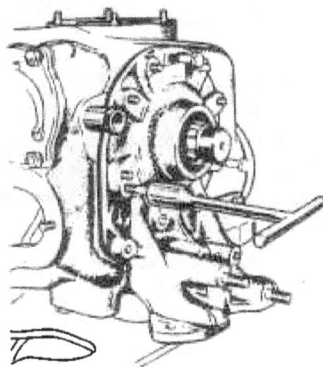

FIG. F5.   REMOVAL OF REAR COVER.

SILVER WRAITH — SILVER DAWN — BENTLEY MK. VI.

R. TYPE BENTLEY — PHANTOM IV.

(ix) Remove hexagon headed adaptor (plug) at rear of gearbox, then remove shock damper oil pump and speedometer drive unit. See Fig. F8.

(x) Next, remove second motion shaft -
Remove locating screw from underneath rear end of gearbox. Insert extractor into front end of shaft and withdraw. Tapped hole in shaft, $\frac{3}{8}$" x 20 T.P.I. Lift or lever shaft out of position and collect distance washers.

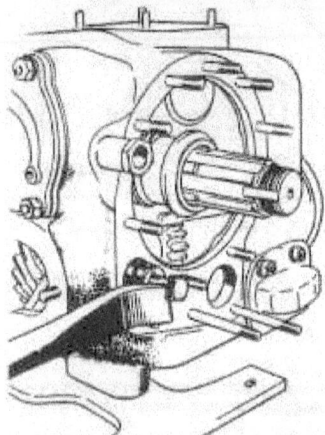

FIG. F6. EXTRACTING LOCATING PIECE.    FIG. F7. REMOVING SERVO CROSS-SHAFT.

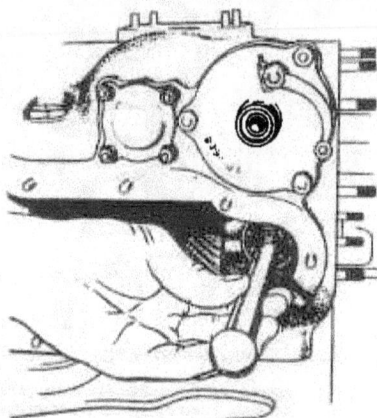

(xi) Remove guide shaft for the operating forks, extractor hole in end of shaft, $\frac{1}{4}$" x 26 T.P.I. See Fig. F9.

Mark the two operating forks to allow re-assembly in original positions. Put reverse motion shaft in forward position.

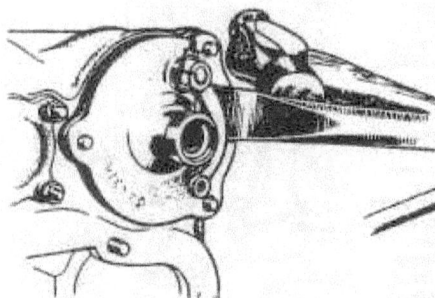

FIG. F8. REMOVING DAMPER PUMP.    FIG. F9. REMOVING SELECTOR FORK GUIDE SHAFT.

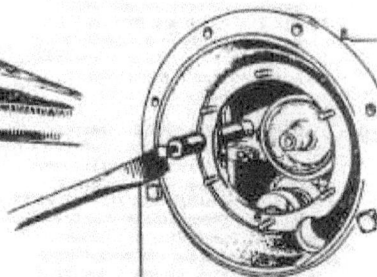

(xii)  Secure extractor 1639/T1008 to front end of gearbox and screw threaded
end of spindle into nose of third motion shaft. Place circular guide
sleeve into rear end bore of gearbox (with recess in sleeve facing
inwards). Operate to push shaft rearwards until central ball bearing
(6, Fig. F10), is just clear of its housing. Remove tool, and manoeuvre
the two operating forks from the shaft assembly and withdraw shaft and
forks from box. Temporarily fit Jubilee clips to rear end of third
motion shaft.

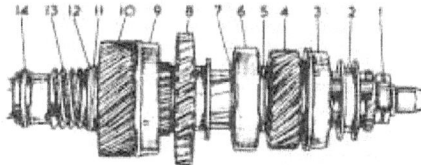

FIG. F10. THIRD MOTION SHAFT.

| | |
|---|---|
| 1. Retaining circlip. | 8. First speed gear. |
| 2. Sliding piece. | 9. Cone assembly. |
| 3. Cone assembly. | 10. Second speed gear. |
| 4. Third speed gear. | 11. Bush second speed gear. |
| 5. Bush third speed gear. | 12. Adjusting washer. |
| 6. Ball bearing. | 13. Servo drive worm gear. |
| 7. Adjusting washer. | 14. Adjusting washer. |

(xiii) Remove the reverse
motion shaft, normally
this is not necessary.
See Fig. F11.

Remove locating screw
from underneath front
end of gearbox. Insert
extraction into rear end
of bearing shaft, (ex-
traction hole 3" x 20 TPI)
and while holding reverse
motion shaft, remove the
bearing shaft. Collect
the thrust washer.

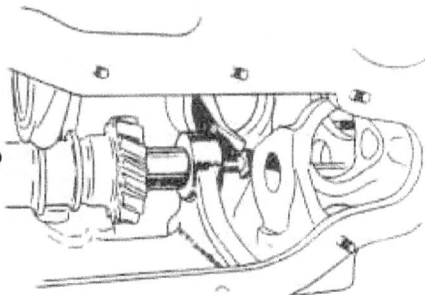

(xiv) To remove the selector
shafts.

NOTE:- Selector shafts cannot be       FIG. F11. REMOVING REVERSE MOTION SHAFT.
removed without first
removing reverse motion shaft.

(xv)  Remove eccentric pivot pin (3, Fig. F12) after removing retaining plate 1.
Withdraw eccentric pin with pliers, then remove actuating lever, 4.

(xvi) Remove the eccentric pivot pins 3 and 7, Fig. F13, and the actuating
levers 5 and 9 and the adjusting washer, if fitted.

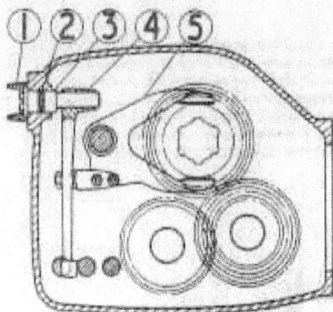

1. Plate, pivot pin.

3. Eccentric pivot pin.

4. Actuating lever.

5. Operating fork.

FIG. F12.   THIRD AND FOURTH SPEED
            OPERATING LEVER ASSEMBLY.

(xvii)   Invert gearbox and remove bottom cover plate, (spring loaded), collect
         the three springs and the ½" dia. balls.  Remove cover from rear of
         gearbox.

(xviii)  Unlock jaws on selector shafts, two on reverse shaft and on each on first
         and third shafts.  Remove and withdraw shafts.

(xix)    Insert a piece of suitably bent wire into outer rear bore of gearbox
         carrying rear end of third and fourth speed selector shaft, press the
         inter-locking ball into the centre bore and recover ball.  Insert wire
         into bore nearest centre of gearbox and carry out same operation to
         recover ball.  These two balls are 5/16" dia.

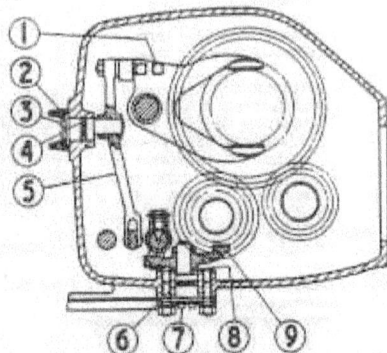

1. Operating fork.

3. Eccentric pivot pin.

5. Actuating lever.

7. Eccentric pivot pin.

9. Actuating lever.

FIG. F13.   FIRST AND SECOND SPEED
            OPERATING LEVER AND REVERSE
            ACTUATING LEVER ASSEMBLIES.

DISMANTLING AND RE-ASSEMBLY - 3rd Motion Shaft:

Dismantling:

1. Mount shaft horizontally in vice, front end upwards, and remove circlip.

2. Press down front ends of the two spring loaded keys (31, Fig.F20) and slide the sliding piece forward one inch, press down rear end of keys and remove sliding. piece from shaft. Collect keys, four balls and two springs.

NOTE:- Current type shafts are fitted with four small plungers in place of the four balls, together with a modified type of sliding piece, keys and spring as in paragraph 5.

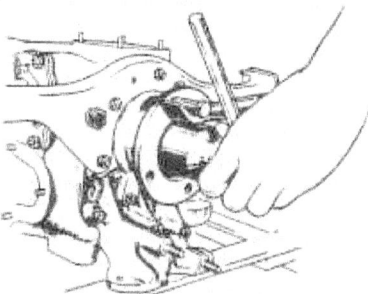

FIG. F14.   REMOVAL OR ASSEMBLY OF RETAINING NUT OF 3RD MOTION SHAFT.

3. Insert pointed end of stiff wire into one of the small holes in the 3rd speed driven gear, revolve gear to locate head of spring loaded pin (27). Depress pin, and with small screw driver, push locating key (13) inwards, remove wire and push key inwards to disengage with internally splined thrust washer (29). Rotate thrust washer to line up with shaft splines and withdraw.

4. Remove gear (26) complete with two floating bushes, wire bushes to gear for correct re-assembly.

5. Remove sleeve (25) carrying ball bearing (10). Prevent loss of spring loaded pin and spring; with hole in sleeve facing up, slide sleeve forward an inch. Place thumb on shaft in line with hole, slide sleeve forward to release pin and trap with thumb. Remove pin, spring and adjusting washer.

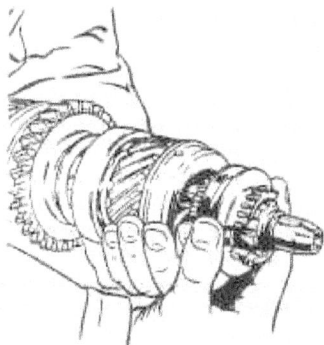

FIG. F15.   REMOVAL OR ASSEMBLY OF 3RD AND 4TH GEAR SLIDING PIECE.

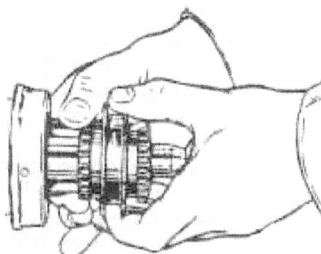

FIG. F16.   DISMANTLING OR ASSEMBLY OF 3RD AND TOP SLIDING PIECE ON 3RD MOTION SHAFT.

6. Remove all parts from rear end of shaft up to 1st speed driven gear. Slide the two keys (22) towards the rear end and remove. With two screw drivers, prise up the two spring loaded balls (24) and slide gear off shaft.

Assembly:

NOTE: On chassis previous to B-159-DZ the 3rd motion shaft has two grooved keyway splines at the front end, and it is essential that these are subjected to a crack test in the area where the two front cones contact the grooved splines.

The magnetic test is the most reliable, but if this not possible, a chalk test should be carried out.

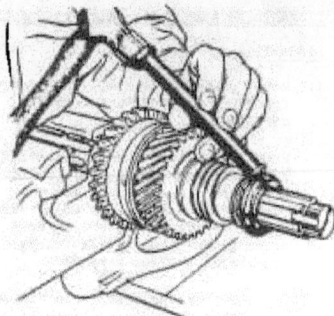

FIG. F17. REMOVAL OF INNER RING OF BALL RACE ON 3rd MOTION SHAFT.

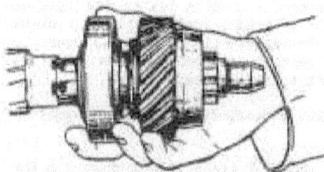

FIG. F18. DISMANTLING OR ASSEMBLY OF 3rd SPEED GEAR AND CENTRE BEARING ON 3rd MOTION SHAFT.

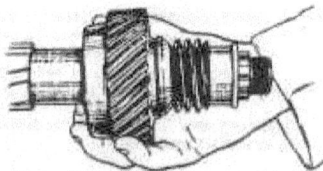

FIG. F19. REMOVAL OR ASSEMBLY OF 2nd SPEED GEAR AND SERVO WORM DRIVE ON 3rd MOTION SHAFT.

If a crack is discovered, a new later type shaft with strengthened (non keyway) splines and associated parts must be fitted. These shafts have been incorporated in all gearboxes after B-159-DZ.

Also, on all gearboxes, from chassis B-159-DZ, square edged type ball bearings have been fitted to the rear ends of the 3rd motion shaft and the 1st motion shaft. This modification must be incorporated in all earlier boxes whenever the box is dismantled for any purpose. The square edge must always be fitted so that it is adjacent to the spring retaining ring.

Inspection:

Check that the two flat keys (22) are of equal length, hold together with the groove of each key in line with each other, grind or stone off as necessary.

Check the three cones, place each cone on respective gear and check that they can be lifted off quite freely, polish as necessary, but do not touch bronze insert.

Check all keys are free from burrs and that they slide easily in the grooves of the shaft.

SILVER WRAITH — SILVER DAWN — BENTLEY MK. VI.

R. TYPE BENTLEY — PHANTOM IV.

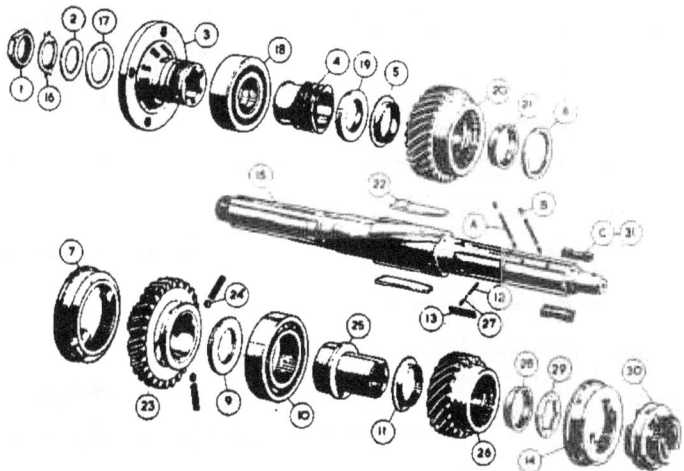

FIG. F20. 3RD MOTION SHAFT.

1.  Nut - Rear End.
2.  Washer - Rear End.
3.  Coupling Flange.
4.  Worm Gear - Servo Drive.
5.  Bush - 2nd Speed Gear.
6.  Adjusting Washer (range off).
7.  Cone 2nd Speed Gear.
9.  Adjusting Washer (range off).
10. Ball Bearing.
11. Bush - 3rd Speed Gear.
12. Spring - Pin.
13. Key, locating.
14. Cone, 3rd Speed Gear.
15. Shaft, 3rd Motion.
16. Lockwasher - Rear End.
17. Adjusting Washer (range off).

18. Ball Bearing - Rear End.
19. Adjusting Washer (range off).
20. Gear, 2nd Speed.
21. Bush - 2nd Speed Gear.
22. Key - 1st Speed Gear.
23. Gear, 1st Speed.
24. Ball (2 off) 1st Speed Gear.
25. Sleeve.
26. Gear, 3rd Speed.
27. Pin - Locating Key.
28. Bush - 3rd Speed Gear.
29. Thrust Washer.
30. Sliding Piece.
C.  Key (2 off).
B.  Plunger (4 off).
A.  Spring (2 off).

On the early type sliding piece (30) there are two non-full teeth, at the rear end, i.e. the end which has the least number of teeth and which engages with the 3rd speed gear. If not already removed, these two teeth must be removed by grinding. If a new strengthened spline shaft is to be fitted, this operation will not be necessary as the existing sliding piece is replaced by the later Type.

1. Mount the rear end of the shaft in vice with hole for pin (27) uppermost. Fit the adjusting washer (9) (note, internal chamfer on abutting face) followed by the splined sleeve (25) carrying the ball bearing (10). Fit floating bush (28) i.e. the one with the small diameter flange, to the front end of the 3rd speed driven gear (26). Fit the rear bush (11), do not oil bushes at this stage. Place gear on the splined sleeve, refit internally splined thrust washer (29) and insert locating key (13) about half way to prevent washer turning.

2. The gear should be fitted so that it rotates freely without measurable end float.
   Should the bushes show signs of wear which would allow any tilting of the gear, new bushes must be fitted with, if necessary, new adjusting washers.

   If a new 3rd or 2nd speed driven gear is to be fitted, then new bushes must be fitted with the gear.

3. Remove splined thrust washer, 3rd speed gear and splined sleeve and oil bushes of gear. Fit pin (27) to spring and replace in shaft. Press down pin and slide sleeve onto shaft, ensuring small hole registers with pin. Refit 3rd speed gear and thrust washer. Insert locating key (13), (hole towards rear of shaft and counterbore facing innermost), into a spline of the thrust washer and keyway in the sleeve and slide up to pin.

   Insert pointed end of stiff wire into small hole in gear to depress pin and slide pin inwards to meet wire. Remove wire and slide key inwards until head of pin fully engages and locks with counterbore in key.

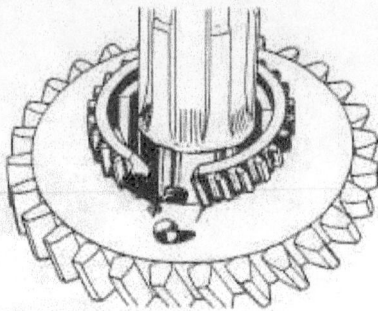

FIG.F21.   ASSEMBLY OF BALL AND PLUNGER, 1st SPEED GEAR, ON 3rd MOTION SHAFT.

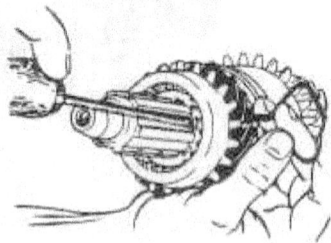

FIG.F22.   ASSEMBLY OF KEY WASHER, 3rd SPEED GEAR, on 3rd MOTION SHAFT.

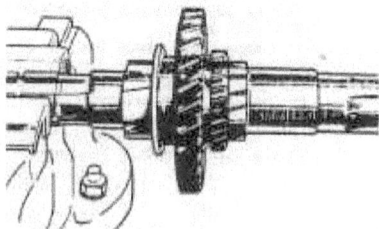

FIG.F23.   1st SPEED DRIVEN GEAR.

# SERVICE HANDBOOK

### SILVER WRAITH — SILVER DAWN — BENTLEY MK. VI.
### R. TYPE BENTLEY — PHANTOM IV.

Next, check end float of 2nd speed gear. Fit adjusting washer (6) on shaft (note, internal chamfer on abutting face). Fit floating bush (21) to front end of 2nd speed gear. Fit the rear bush (5), but do not oil bushes.

Fit gear to shaft and place adjusting washer (19) correctly in position. Push hard against washer and check with feelers amount of end float of gear. If in excess of .002" fit thicker adjusting washer. The gear should rotate freely with minimum end float.

Remove gear and washer and oil bushes. Place cone (7) on shaft followed by washer (6), gear (20), washer (19), worm gear (4) the two halves of inner race for ball bearing (18), coupling flange (3) and adjusting washer (17).

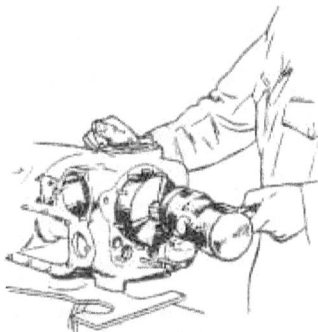

FIG. F24. POSITIONING 3rd MOTION SHAFT.

Push hard against coupling flange and check adjusting washer (17) stands out, beyond shaft shoulders from .006" - .010", to give required nip to coupling flange.

6.  If a new 3rd motion shaft is to be fitted to an early type gearbox, this will be to the latest pattern with strengthened splines and will require. a new sliding piece for the 3rd and 4th speed gear also new keys, plungers and springs. Fit as below.

    (a) With the cone (4), in position, place new sliding piece in position correct way round, slowly rotate 3rd speed gear and check each tooth of sliding piece in every position in the gear for freedom of engagement, then remove sliding piece.

    (b) Fit the two springs (A, Fig.F20) to the two holes in the front end of the shaft, fit a plunger (B) to the end of each spring and fit bridge shaped keys (C) over plungers. Compress keys and slide sliding piece into position.

    Temporarily fit Jubilee clip to front end of shaft for retention of parts.

## RE-ASSEMBLING:GEARBOX:

On certain cars prior to Chassis B-170-BH, the spring retaining ring groove in the front cover was machined 4.012" + .01" in diameter, and this should be machined out to 4.074" + .01" and a larger ring fitted unless square edged bearings have been installed.

1.  Replace selector shafts and jaws.
    Insert reverse selector shaft and thread on jaw, then the reverse actuating jaw, and lock into position, see Fig.F13. With the shaft in approximately neutral position, place one of the two .3125" dia. balls in centre of bore and manoeuvre until it engages with spherical recess near end of reverse shaft.

# SERVICE HANDBOOK

### SILVER WRAITH — SILVER DAWN — BENTLEY MK. VI.
### R. TYPE BENTLEY — PHANTOM IV.

Insert 1st and 2nd speed selector shaft and fasten jaws as above. Line up jaw with selector jaw on reverse shaft and insert .3125" ball as above.

Insert 3rd and 4th speed selector shaft and fasten selector jaw, position shaft to line up jaw with others. See Fig. F12.

Replace the three .500" dia., balls and springs, refit bottom cover plate and rear end cover.

Replace adjusting washer (8, Fig. F13) and actuating lever (9), engage latter in reverse actuating jaw. Fit eccentric pivot pin (7) and ensure correct positioning.

2.  **Replace Reverse Motion Shaft:**
Hold reverse shaft in gearbox and enter bearing shaft into bore of rear support and then reverse shaft. Place bronze thrust washer on front end of shaft with grooved side against gear. Line up hole in front end with corresponding hole beneath gearbox and fit locating screw and lock.

Place reverse selector shaft in neutral position and insert .092" feeler between bearing shaft support rear end and face of small gear on shaft, rotate eccentric pin, beneath gearbox, until light nip is felt and then lock pin in position.

Move reverse motion shaft and check selector shaft for "over-ride" i.e., amount by which the selector shaft can be moved beyond position when gear is fully engaged. Over-ride of approximately 1/16" is necessary to ensure ball locates in groove on shaft and thus prevents gear from jumping out of position.

3.  **Replace 3rd Motion Shaft:**
With the actuating levers on their respective pivot pins, place the two forks loosely in position. Grease centre ball bearing and remove Jubilee clip from rear end and enter shaft assembly into rear end of gearbox. Do not enter centre ball bearing in housing at this stage.

Partially engage rear fork in groove of 1st speed driven gear and manoeuvre front fork into groove of sliding piece. Line up centre ball bearing, hold shaft in position and tap end of shaft until bearing is in position, simultaneously guiding forks into corresponding grooves. See Fig. F24.

Connect forks to respective actuating levers and refit guide shaft; guide shaft should slide easily into position.

4.  **Replace 2nd Motion Shaft:**
Place the two roller bearings in position and insert shaft into box. To correctly position case hardened washer at rear end, a mandrel should be used for support.
Fit an adjusting washer to front end of shaft that will allow it to be a light tap-in fit so as to pre-load the shaft .002" - .004". Fig.F26.

Fit the bearing shaft and line up hole in rear end with corresponding hole beneath gearbox, fit locating screw and lock. See Fig. F25.

5.  **Replace Servo-Drive Assembly:**
Tap servo drive shaft, minus housing, into gearbox, simultaneously rotating 3rd motion shaft to allow engagement of gears, when midway, fit locating piece. Manoeuvre top of oil-seal onto shaft, turn housing until drain hole is at the bottom and secure in place. See Fig. F27.

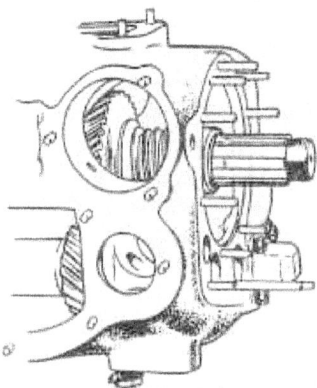

FIG. F25. LOCATING 2nd MOTION SHAFT.

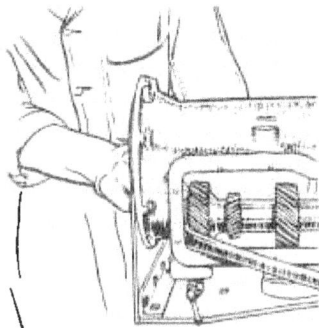

FIG. F26. SETTING ADJUSTING WASHER
2nd MOTION SHAFT.

6.   Replace Damper Pump Unit.
Remove Jubilee clip from rear end of 3rd motion shaft and place cover in
position, before securing, replace coupling flange and tap up coupling to
close gap between the two halves of the inner race of the ball bearing.
Place torque bracket in position and secure cover to gearbox.
Place adjusting washer, lock washer and nut on 3rd motion shaft, tighten
and lock.

Replace casing of damper pump unit, rotating speedometer drive shaft to
engage gears. Secure in position.

Replace hexagon headed adaptor
plug, using jointing compound,
into rear of gearbox.

Remove Jubilee clip from front
end of the 3rd motion shaft and rotate to
bring the two keyways horizontal. See
Fig. F28, but rotate to correct position.
Place cone on front end of shaft ensuring
driving torques are in line with keys.
Grease and place rollers in position in
1st motion shaft and hold with Bakelite
retainer. Rotate front end cover to bring
drain hole at bottom and enter 1st motion
shaft on to nose of 3rd motion shaft.
Secure front end cover.

FINAL CHECKING:

3rd and 4th speed cones - place
3rd and 4th speed selector shaft in
neutral and rotate 3rd motion shaft to
bring keyway opposite side cover opening,
with feeler measure clearance between end
face of the stop in the key (i.e. end face of
portion projecting above grooved spline)

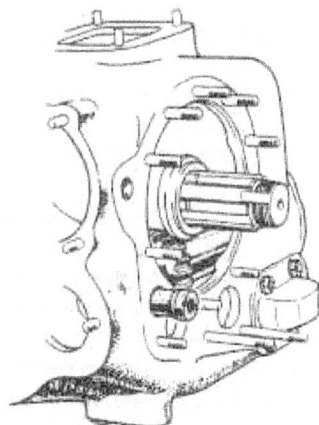

FIG.F27. LOCATING PIECE FOR SERVO SHAFT.

- F.15 -

# SERVICE HANDBOOK

### SILVER WRAITH — SILVER DAWN — BENTLEY MK. VI.
### R. TYPE BENTLEY — PHANTOM IV.

and the rear face of the cone. The later type shafts fitted with bridge shaped keys and strengthened splines are not stepped, therefore measurement is taken from extreme end face of keys.

Leave feeler in position, then hold opposite cone squarely on the taper of the 3rd speed driven gear and with feeler measure clearance.

Each cone should have clearance of .020" - .030"; to equalise, adjust eccentric pin, R.H. side (front) of gearbox, then recheck "over-ride".

2nd speed cone - place 1st and 2nd speed selector shaft in neutral. Hold 2nd speed cone squarely on taper of 2nd speed gear and measure clearance between end face of flat key and rear face of cone. This should be .015" - .025", adjust by eccentric pin, R.H. side (rear) of gearbox. Check "over-ride" of selector shaft, a small adjustment can be made with eccentric pin ( ).

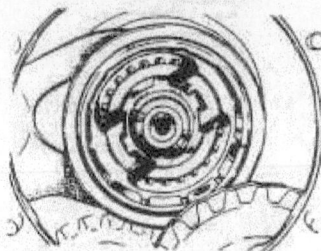

FIG. F28. PRIOR TO FITTING 1st MOTION SHAFT. (Rotate to correct position).

Refit damper pump oil feed tube assembly, dipstick, side cover plate and drain plug.

Using "Wellseal" jointing compound, refit bearing covers of selector lever shaft assembly, ensure line up of oil return holes. Fit top cover.

## GEAR CHANGE CONTROLS

### SIDE CHANGE MECHANISM:

After refitting gearbox to chassis and coupling selector lever and change gear lever shafts at the spherical joint, and before refilling gearbox, check as under:-

(i)   Remove gearbox top cover, and with gear lever in neutral, check that end of selector lever (1) Fig. F29 is central in jaw (2) of 3rd and 4th speed selector shaft.

(ii)  Next, check that there is a small gap at "A" Fig. F30, when gear lever is moved into each gear position. Any difference between front and rear gaps, adjust by pivoting gate forward or rearward, this should also give reasonably equal clearance at gaps "B" and "C"

(iii) Check gear lever slides freely from end to end of gate. It should be noted that if adjustment of tie rod, Fig. F31, has been altered, this may cause cross binding of gear lever when moving

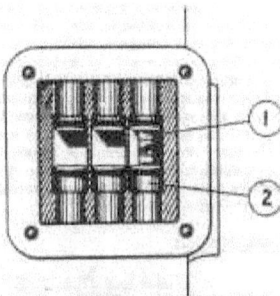

FIG. F29.   CORRECT POSITION OF END OF SELECTOR LEVER (NO.1) IN RELATION TO JAW.

SILVER WRAITH — SILVER DAWN — BENTLEY MK. VI.

R. TYPE BENTLEY — PHANTOM IV.

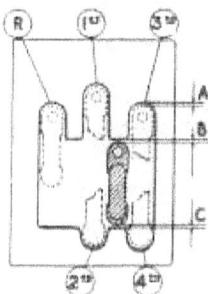

FIG.F30. PLAN VIEW OF GATE.

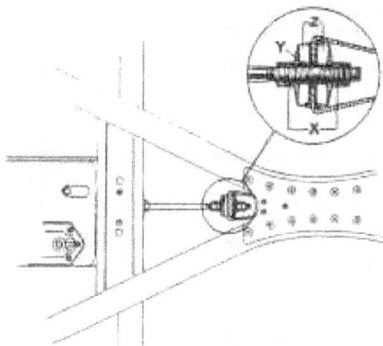

FIG.F31. TIE ROD - ENGINE FORE AND AFT LOCATION.

it across gate and/or prevent free movement of selector lever (1) Fig.F29, across the jaws.

The tie rod locates the engine and gearbox unit in a fore and aft direction and the adjustment should be such as to impose no fore and aft deflection in the rubber mounting block on which the rear end of gearbox rests.

If new rubber blocks have been fitted, adjust front end of tie rod on bench, and with the rear end parts loosely fitted, refit assembly:-

Front end tie rod - with the inner nut screwed back as far as possible, tighten adjusting nut at front end until distance piece between the two rubber blocks is clamped. Tighten locknut.

Rear end tie rod - screw up adjusting nuts "X" Fig. F31, finger tight, and then tighten equally until distance piece "Y" is clamped; which will equally preload the rubber blocks "Z". Tighten lock nuts.

Removal and Dismantling:

(i)     Remove bolts and uncouple gear lever shaft 4, Fig. F32, from selector lever shaft at spherical joint. Mark relationship of flanges and collect spring.

(ii)    Disconnect reverse light switch. Disconnect and remove gate bracket from frame, leave packing pieces in position. Packing pieces are .036" thick and are for adjustment of change gear lever shaft with selector shaft, note that in certain cases tapered pieces may have been fitted.

(iii)   Remove cover (10), pinch bolt (12), take care of small spring link fitted on bolt. Remove lever from shaft. Remove key (11), ferrule (15) washer, and spring (17). Remove shaft complete with rubber stop (20) and then the gate (8) from the bracket.

(iv)    Next, remove ball bearing assembly (6) and spring (18). Tap out washer (19) complete with outer bearing race (5).

# SERVICE HANDBOOK

### SILVER WRAITH — SILVER DAWN — BENTLEY MK. VI.
### R. TYPE BENTLEY — PHANTOM IV.

NOTE:- The rubber stop (20) acts
as a cushion preventing
change gear lever contacting
right-hand side of gate when
changing from 3rd to 4th speed.

Re-Assembly:

Lightly grease the ball
bearing assembly and re-
assemble the parts in the
reverse order of dismant-
ling. The two springs
(17 and 18) are identical.
Check position of gate.

COLUMN GEAR CHANGE:

On and after chassis
Bentley B-281-LGT, Silver Dawn
LSCA-11, and Silver Wraith LWHD-78
key hole type sockets are fitted
to the gear change cross tube
assemblies and the gear change pull
rod assemblies, 5 in all. The type

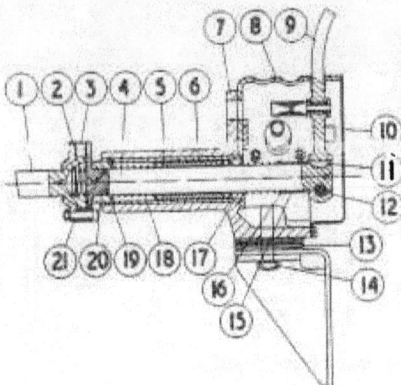

FIG. F32. SECTION - SIDE CHANGE GEAR MECHANISM.

fitted can easily be verified by noting, on early type the adjusting plug is operated
by a screw-driver whereas on the later type it is by means of a spanner.

Cases of rattle from the steering column reported on early models can be
cured by the fitting of two rubber bushes on the inner tube, one about half way up
and the other at the lower end.

Re-Assembling:

The two gear change cross tubes are connected to the ball pins of the outer
lever, (on the left-hand side of the gearbox) attached to the horizontal selector
lever of the gearbox. The other ends are connected to the upper and lower ball pins
of the gear change bellcrank lever, the upper end being connected to the lever fitted
at the lower end of the gear change tube attached to the steering column. The ball
joints of the pull rods and cross tubes are fitted with rubber dust covers.

(i)   Roughly assemble the entire linkage. Refer to Fig.F33, jaw X.17 and jaw 11
      should be fully screwed home on rods 28 and 30. The adjusting rods at the
      other ends of these rods should be screwed on for ¼" temporarily. These
      rods should be as near as possible to their maximum length.

(ii)  Adjust the ball joints of rods 42 and one end of 29, so that all end play
      is removed, but without producing stiffness.

(iii) Adjust jaw 8, so that the gear control lever lies just above the
      horizontal when in the neutral position.

(iv)  Remove the gearbox cover and observe the position of lever 19.

(v)   With the gearbox in neutral, check the travel of lever 19 across the
      gate. In the 3rd and 4th positions, it should be just clear of the
      side of the gate.

SILVER WRAITH — SILVER DAWN — BENTLEY MK. VI.

R. TYPE BENTLEY — PHANTOM IV.

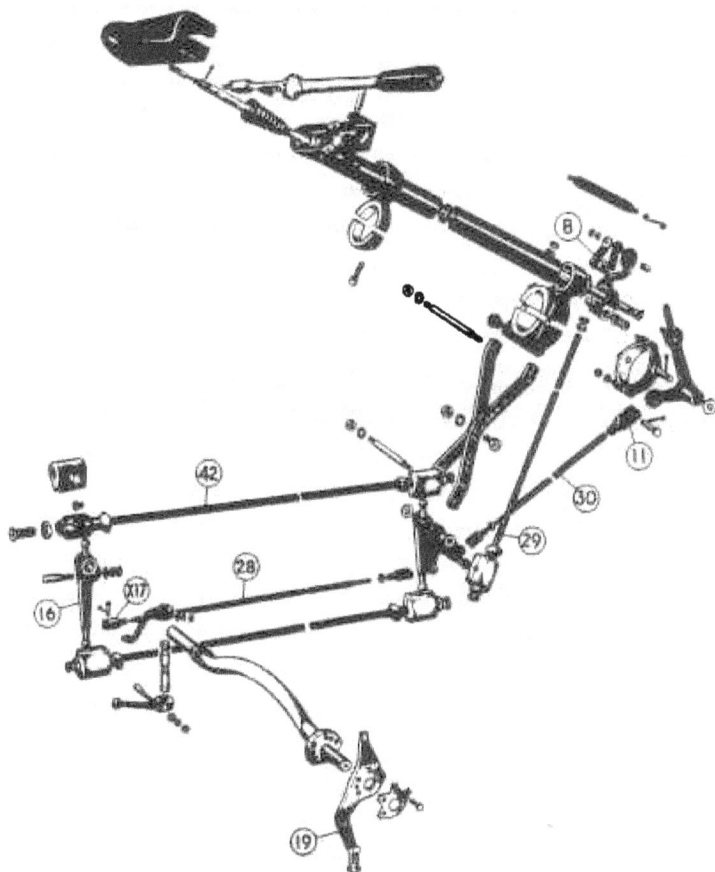

FIG. F33. COLUMN GEAR CHANGE MECHANISM.

If necessary, slacken the nut and loosen the cotter retaining the lever 16, and tap it along the shaft until the rubber stop between the lever 16 and the side cover of the gearbox make contact just before the lever fouls the side of the gate.

(vi)    Next, lift the control lever into 1st and 2nd position, i.e. against the reverse stop, and check the travel of lever 19 across the gate. This should be such that the 1st and 2nd gears can be engaged without lever 19 fouling the side of the gate.

(vii)    If the lever 19 does not move far enough to give the correct alignment with the gate, shorten the rods 28 and 30 equally by screwing up the jaws until the adjustment is correct.

(viii)    If the lever 19 moves too far to align with the gate, lengthen the rod 30 by fitting a 5/16" nut (L.H. thread) behind the jaw 11, and adjust as necessary.

(ix)    Replace the gearbox cover.

(x)    Split pin all joints and check locknuts.

When checking the settings on a complete car, so as to avoid removing the floor boards, it is permissible to set rods 28 and 30 by putting the gearbox in 1st or 2nd, holding the gear lever up against the reverse stop and adjusting the rod lengths until the pins can just be fitted.

SILVER WRAITH — SILVER DAWN — BENTLEY MK. VI.

R. TYPE BENTLEY — PHANTOM IV.

SECTION F.

GEARBOX

(AUTOMATIC TYPE)

Gear Ratios:    1st speed  3.8195:1
                   2nd speed  2.6341:1
                   3rd speed  1.450 :1
                   4th speed  Direct
                   Reverse    4.3045:1

Fluid Capacity  20 pints (Imperial)
Weight          250 lbs. approx.

GENERAL:

The automatic gearbox consists of the following power units, the flywheel cover; the front planetary unit, consisting of single reduction planetary gears; the fluid coupling or torus members; the rear planetary unit, comprising compound planetary gears; and the reverse unit.

The planetary units each consist of three planet gears meshed between a sun gear and an internal gear, which is integral with a drum and can be held from rotating by a band. The band is applied or released by a servo mechanism, consisting of a double-acting piston and cylinder operated by oil pressure.

In addition to the power units, there are the hydraulic control parts, two oil pumps, governor and control valves, and the piston and servos which operate the front and rear clutches and bands by means of which the various gear ratios are obtained.

Fluid (oil) provides the transmission medium and gear changes are effected automatically by the control valve unit, operated by oil pressure. The road speeds at which the gear changes occur, vary according to the demand for power from the accelerator pedal.

Because the fluid in the unit performs several entirely different functions, transmitting the drive, changing the gear ratios, and lubricating the mechanism, a special fluid must be used.

OPERATION:

The power flows from the flywheel and flywheel cover to the front planetary unit, then forward to the fluid coupling, then back through the main shaft to the rear planetary unit, through the reverse unit (which is idling in forward speeds) to the output shaft.

SILVER WRAITH — SILVER DAWN — BENTLEY MK. VI.

R. TYPE BENTLEY — PHANTOM IV.

| Unit in OPERATION. | NEUTRAL | 1st | 2nd | 3rd | 4th | Rev. |
|---|---|---|---|---|---|---|
| Front Band | - | X | - | X | - | X |
| Front Clutch | - | - | X | - | X | - |
| Rear Band | - | X | X | - | - | - |
| Rear Clutch | - | - | - | X | X | - |
| Reverse Anchor | - | - | - | - | - | Engaged |

In first speed, both bands are on and both units are in gear to provide maximum gear reduction.   In second speed, the front band is off and the front clutch is engaged so that the front unit is in direct drive and only the reduction of the rear unit is effective.   In third speed, the front band is re-applied and the front clutch released so that the front unit is again in reduction, but the rear band is released and the rear clutch is applied, so that the rear unit is in direct drive.   In fourth speed, both clutches are applied and both bands released, which means that both units are in direct drive.

In reverse, the front unit is in gear, the rear unit idling, and the reverse anchor engaged.   This causes a further reduction and change in direction of rotation of the output shaft.

In neutral (engine running) all bands and clutches are released and the mechanism idles.

The two oil pumps provide the operating pressure necessary for control of the units in the gearbox.   The front pump is driven by the crankshaft and therefore operates at engine speed and only when the engine is running.   The main duties of the front pump are:-

1.   To provide main oil pressure in the control valve unit for operation of the servos and clutches.

2.   To supply boosted pressure when required.

3.   To maintain the pressure in the fluid coupling.

The rear pump is mounted with the governor on a shaft driven from the output shaft of the gearbox, and the oil delivery from this pump is therefore proportional to the road speed of the car.

When the engine is started, the front pump builds up pressure in the regulator valve housing, and the fluid is directed through channels to the valve body, and after the pressure has been built up to a pre-determined figure - to the fluid coupling and the lubrication system, by means of the pressure regulator valve and a spring loaded shuttle valve.

As the car moves forward, the rear pump operates to build up pressure against the shuttle valve.   When the pressure reaches a pre-determined amount, it opens the shuttle valve so that the rear pump provides all the pressure to the valve body.   The front pump then becomes a low pressure lubricating pump to supply fluid to the fluid coupling and the lubricating system.

### SILVER WRAITH — SILVER DAWN — BENTLEY MK. VI.
### R. TYPE BENTLEY — PHANTOM IV.

With the control unit being operated by the pressure from the rear pump and therefore by the vehicle's road speeds, up-shift and down-shift of the gear ratios would always be at the same vehicle speed. In order to delay the shifts when high acceleration is required, the accelerator pedal is linked to a valve (throttle valve) in the control valve unit.

By pressing down on the accelerator pedal, to operate the valve, main oil pressure is allowed to modulate against the governor oil pressure by means of regulator valves incorporated with each of the three shift valves.

The regulating pressure has the effect of increasing the strength of the shift valve springs, and therefore if the driver requires quick acceleration and puts his foot down, the excess use of the accelerator in relation to the speed of the car, gives up-shift at a higher road speed, giving better acceleration by the use of the lower gears.

| Up-Shift | Light Throttle | Full Throttle |
|----------|----------------|---------------|
| 1st - 2nd | 6 m.p.h. | 15 m.p.h. |
| 2nd - 3rd | 11 m.p.h. | 35 m.p.h. |
| 3rd - 4th | 20 m.p.h. | 65 m.p.h. |

Extreme pressure on the accelerator pedal, beyond the normal limit of its travel, actuates a plug operating together with the throttle valve, which gives forced down-shift for "kick-down" when required.

Pressure from the throttle valve also acts proportionally against a compensator valve which feeds boosted pressure to the brake band servos and clutch pistons as engine speed increases, thus giving increased pressure to compensate for additional engine torque.

The manual valve, operated by the lever on the steering column, is a simple valve which closes or opens passages in the control valve unit to obtain the conditions required when the lever is set in the N. 4. 3. 2 and R positions.

#### COASTING OR TOWING:

Coasting or "freewheeling" down hills with the engine switched off, must definitely be avoided, as this is likely to cause severe damage to the Automatic Gearbox mechanism. This damage can occur with the manual control lever in any of the five positions, including position "N".

When the engine is switched off, the front oil pump becomes inoperative, lubrication is reduced, and the rear servo brake band applied, and under these circumstances, the application of the rear band causes excessive over-driving of the front unit clutch plates, with the probability that these clutch plates will burn out, necessitating a major gearbox overhaul.

Similarly, if, in the event of an accident, it should become necessary to tow a car fitted with the Automatic Gearbox, the following precautions must be observed before the car is moved.

1.  Make certain that the gearbox was operating satisfactorily until the time of towing. If there is any sign of mechanical failure or breakage in the gearbox, the car must be transported.

2.  Release the lock nut and slacken off the rear band adjusting
    screw 4½ complete turns.  Re-tighten the lock nut.

3.  Keep the control lever in "N" throughout and maintain where
    possible, a towing speed between 15 and 25 miles per hour.
    Distances must at all times be kept to a minimum.

IMPORTANT:

At no time must a speed of 25 miles per hour be exceeded whilst towing.

Towing new cars or cars with new gearboxes, which have covered less than
1,000 miles in service, must be avoided.

TOPPING UP OR CHANGING FLUID:

The Automatic Gearbox should be filled and topped up only with Automatic
Transmission Fluid, Type "A", having an Armour Qualification number prefixed by
AQ/ATF.

Any of the following may be used:-

| | | | |
|---|---|---|---|
| Vacuum Oil Co. ... | Mobiloil Fluid 200 | ... | Type AQ/ATF.101 |
| Shell | ... | Donax T.6 | ... | Type AQ/ATF.103 |
| Wakefield's | ... | Castrol T.Q. | ... | Type AQ/ATF.156 |
| B.P. | ... | Energol A.T.F. | ... | Type AQ/ATF.261 |
| General Motors ... | Hydra-Matic Fluid | | |

Capacity  ...  ...  20 Imp. Pints.

To Drain Gearbox and Fluid Coupling.

It is necessary to remove two drain plugs, one in the gearbox sump and
one in the fluid coupling.

(i)  Clean the area around the sump drain plug and remove plug.

(ii)  Remove the lower bell-housing cover, and if the drain plug
on the fluid coupling is not at the lowest point, turn the
engine by means of the starter motor to bring plug to this
position, and remove the plug.

(iii)  After draining, securely replace both plugs.

To Fill a New Gearbox or, After Complete Draining:

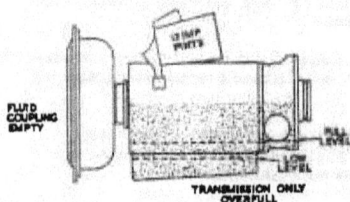

FIG.F.34.  FIRST FILL-IN.

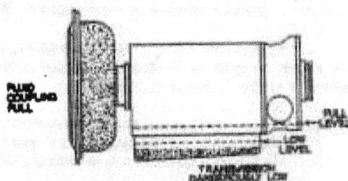

FIG.F.35.  AFTER FAST-IDLE RUN.

SILVER WRAITH — SILVER DAWN — BENTLEY MK. VI.

R. TYPE BENTLEY — PHANTOM IV.

(i)     Remove the dipstick, and through the dipstick filling orifice, add 12 Imp. pints (14 U.S. pints) of a recommended gearbox fluid. See Fig. F.34.

(ii)     Check that the control lever is at "N" (Neutral) and that the hand brake is on, and run the engine at a fast idle for approximately 5 minutes. Switch off the engine. See Fig. F.35.

Add a further 6 Imp. pints (7 U.S. pints). Run the engine at a slow idle and check the fluid level with the dipstick. Add further fluid to bring the level up to the "FULL" mark on the dipstick. See Figs. F.36 and 37.

"LOW" to "FULL" is approximately 2 pints.

NOTE:- Do not overfill; if necessary, drain off to obtain correct level.

Always check oil level with engine warm and running at a slow idle, this allows the fluid coupling to fill and so gives an accurate reading. See Fig. F.37.

### Refilling after Draining Sump only:

(i)     Check sump drain plug is securely tightened.

(ii)     Add 8 Imp. pints (10 U.S. pints) of a recommended gearbox fluid through the dipstick filling orifice.

(iii)     Check that the control lever is at "N" (Neutral) and that the hand brake is on, and run the engine at a fast idle for approximately 3 minutes. Reduce engine speed to a slow idle and check the fluid level with the dipstick. With the engine still running, add further fluid to bring the level up to the "FULL" mark. See "NOTE" above.

### Topping-up Gearbox:

(i)     Set the control lever in Neutral.

(ii)     Run the engine for approximately 3 minutes, and check the fluid level while the engine is still running.

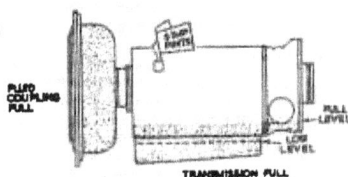

FIG. F.36. SECOND FILL-IN.

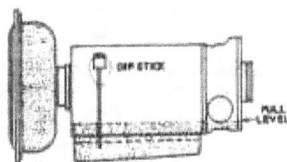

FIG. F.37. CHECKING LEVEL.

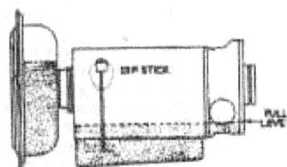

FIG. F.38. INCORRECT READING, FLUID COUPLING NOT FILLED BY ENGINE.

SILVER WRAITH — SILVER DAWN — BENTLEY MK. VI.

R. TYPE BENTLEY — PHANTOM IV.

(iii) Add fluid as necessary, until the level reaches the mark on the dipstick, taking care not to overfill.

THROTTLE AND GEAR CONTROL LINKAGE SETTINGS. (Early Type)

The complete "Factory" settings of the control linkages are given to enable a complete check of all or part of these settings as may be necessary. It should be noted that the final check should be made while the car is on road test.

The gear control linkage should be set first, so that the trapeze which also carries the throttle linkage is correctly positioned. Rod lengths quoted are ball joint or clevis pin centre to centre distance.

(i) Disconnect rod "J".

(ii) Adjust the two horizontal gear control rods to equal lengths such that the trapeze is at right angles to the steering column. Note this length.

(iii) With the gear lever in the "3" position, and the manual lever on the gearbox in the "3" detent position (i.e., second detent back from the forward stop), adjust the vertical gear control rod to suit by means of the jaw at its upper end. Check the other gear lever positions.

(iv) Set rod "A" equal in length to the two horizontal gear control rods.

(v) Set rod "B" to 2.875".

(vi) Set rod "C" to 4.5" on Bentley, 4.875" on Silver Wraith and Silver Dawn.

(vii) With pin "D" held 1.750" from the face of the dash, set lever "E" to lie approximately 2° below the horizontal by slackening the pinch-bolt on lever "F" and rotating lever "E" with its shaft to the desired position.

(viii) With the carburetter butterfly fully closed (idling screw set back and fast idle cam out of action) adjust the face of rod "G" to maintain the 1.750" gauge distance from pin "D" to the face of the dash.

(ix) Holding rod "A" forwards, so that lever "H" is on its off-stop, adjust rod "I" to suit.

(x) Move the linkage towards its full throttle position, and check that when lever "H" reaches its on-stop, the carburetter butterfly also just contacts its on-stop. If not, alter the linkage ratio slightly by slackening the pinch-bolt on lever "F", rotating lever "E" further below the horizontal, and lengthening rod "G" to maintain the 1.750" gauge distance from pin "D" to the face of the dash.

(xi) Set the carburetter idling screw to give 400 r.p.m. hot.

SILVER WRAITH — SILVER DAWN — BENTLEY MK. VI.

R. TYPE BENTLEY — PHANTOM IV.

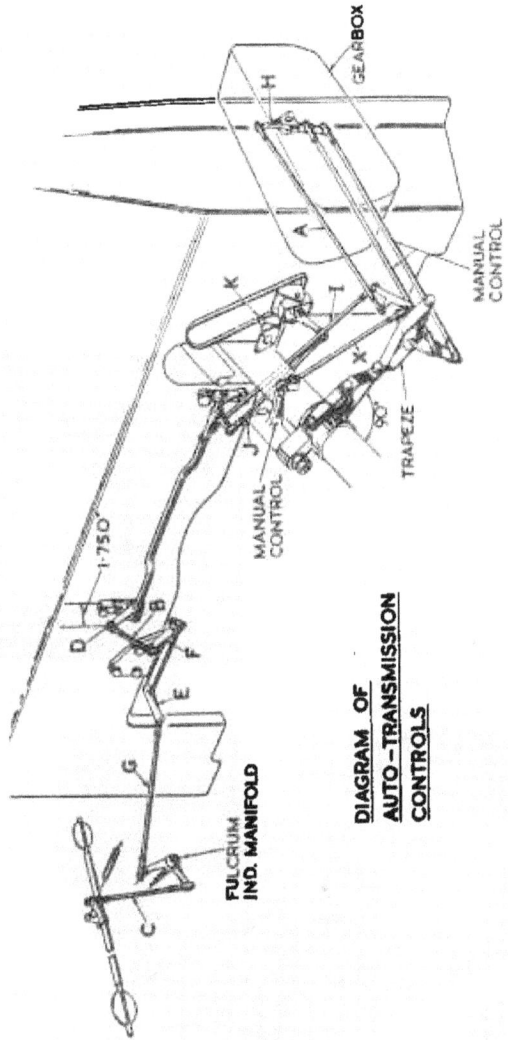

DIAGRAM OF AUTO-TRANSMISSION CONTROLS

FIG. F. 39.   AUTO-TRANSMISSION CONTROLS.

SILVER WRAITH — SILVER DAWN — BENTLEY MK. VI.

R. TYPE BENTLEY — PHANTOM IV.

(xii) Starting with the accelerator pedal in the fully closed position
under the influence of its pull-off spring, adjust rod "J" so
that the pedal is just slightly depressed from this position.
Set stop "K" to contact the pedal in the full throttle position.

(xiii) Check the gearbox operation on the road.   On a level road with
the transmission warm, changes from 1 to 2, 2 to 3, and 3 to 4
should occur at 5-7, 10-13, and 19-22 m.p.h. respectively when
using the minimum throttle opening necessary just to accelerate
the car gradually from rest.

Should the changes occur above these speeds, shorten rod "A"
slightly.   Check the behaviour when making part throttle starts
from rest.   If the changes occur with an excessive amount of
slip (particularly the 2 to 3 change), or if kick-down changes at
full throttle are unobtainable, correct by lengthening rod "A"
slightly.

## 1,000 MILE ADJUSTMENT:

All owners of cars
fitted with the Automatic
Gearbox have been advised
that it is most important to
have the gearbox brake band
adjustment checked after the
first 1,000 miles of initial
running.

1.  Dipstick cover.

2.  Dipstick.

3.  Cover, brake band
adjusters.

4.  Front band adjuster.

5.  Rear band adjuster.

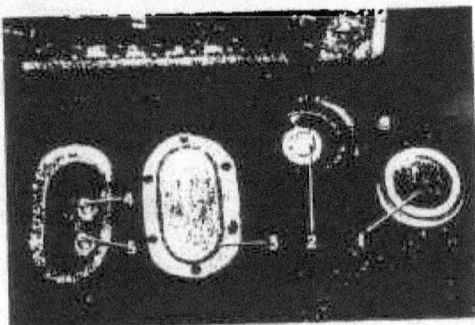

FIG. F.40. ACCESS POINTS IN FLOOR.

The following is the
correct procedure for checking and adjusting the front
and rear brake bands:

Preparation:

(i)     Remove the carpet from
the front interior floor
to expose access cover
over band adjuster
screws.   Remove the
cover, see Fig.F.40.

(ii)    Drive car over pit, or
elevate on hoist.

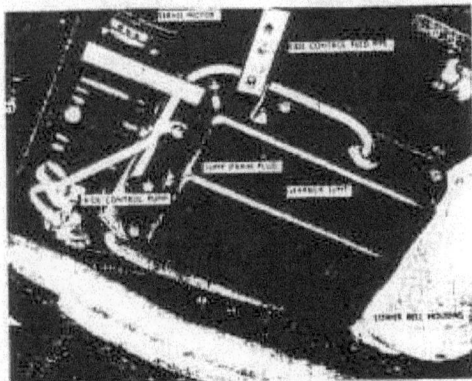

FIG. F.41. GEARBOX - UNDERNEATH VIEW.

(iii) Thoroughly clean the underside of the gearbox sump around the drain plug, sump joint and the oil pipe connection on the ride control pump, see Fig.F.41.

Remove the sump drain plug and drain the fluid into a pan, approximately 10 Imp. pints, or 12 U.S. pints, should drain from the sump.

(iv) Remove the two bolts securing the ride control feed pipe to the ride control pump. see Fig.F.41.

FIG.F.42. GEARBOX - SUMP REMOVED.

Unscrew the twelve securing bolts and remove the sump and sealing joint, taking care not to damage the ride control feed pipe.

Extreme care must be taken to prevent any dirt, dust or lint, etc., from entering the internal portion of the gearbox. Tools and equipment must be perfectly clean.

(v) Remove the oil strainer,Fig.F.42,by pulling the rear end of the strainer down slightly, off the governor feed pipe, and then, sliding bodily rearwards off the main oil pipe.

(vi) Slacken off the locknuts securing the two band adjusting screws.

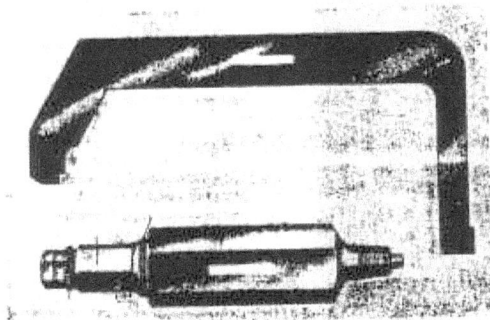

FIG.F.43. GAUGES.

Adjusting the Front Band.

(i) Ensure that the front band is centred on the drum and that the band anchor is seated properly on the adjusting screws.

(ii) Remove plug from front servo body. Adjust Gauge GM.No.J.1693A, Figs. F.43 & 44, by loosening the small hexagon head until approximately $\frac{1}{4}"$ of thread is exposed above the gauge body, and screw the gauge, by hand only, into the plug hole.

SILVER WRAITH — SILVER DAWN — BENTLEY MK. VI.

R. TYPE BENTLEY — PHANTOM IV.

(iii) Tighten the small hexagon head of the tool with the fingers, until the stem of the gauge is just felt to contact the front servo piston. Then, using a wrench, continue tightening the small hexagon head 5½ turns.

(iv) From the top of the gearbox, tighten the front band adjusting screw, Fig.F.40 until the knurled washer on the top of the gauge is just free to turn by hand.

(v) Hold band adjusting screw and securely tighten its locknut.

FIG.F.44. FRONT SERVO - GAUGE IN POSITION.

(vi) Loosen the gauge adjusting hexagon, 6 full turns, and remove. Replace and tighten servo cover plug.

Adjusting the Rear Band:

(i) Place the rear band adjusting Gauge GM.No.J.5071, Figs.F43 & 45, so that the small notch in the short end, rests on the corner of the machined face at the spring end of the rear servo body, and the tip of the long end rests along the servo piston rod.

(ii) From the top of the gearbox, tighten the rear band adjusting screw, Fig.F.40, until the actuating lever just contacts the face of the gauge.

Do not overrun this adjustment, if necessary, loosen the rear adjusting screw two or three turns and repeat the adjustment.

(iii) Hold the band adjusting screw and securely tighten the locknut. Afterwards, recheck with the gauge that the adjustment is correct.

Re-Assembly:

(i) Replace the oil strainer, sliding it first on to the main oil pipe and then firmly positioning the governor feed pipe,

FIG.F.45. REAR SERVO - GAUGE IN POSITION.

in the hole in the top of the strainer.

(ii) Replace the sump using a new sealing gasket, and check that the ride control feed pipe registers approximately on the boss of the pump. Replace and securely tighten sump drain plug and reconnect feed pipe, taking care that it is seating properly when bolts are tightened.

(iii) Replace the cover in the front floor over the adjusting screws, having made sure that the adjusting screw locknuts are securely tightened.

(iv) Refill the gearbox with a recommended transmission fluid.

(a) Add 8 Imp. pints (10 U.S. pints) of transmission fluid through the dipstick filling orifice.

(b) Run the engine with the hand control lever at "N" (Neutral) and the hand brake on, for approximately three minutes, at a fast idle, equivalent to about 20 miles per hour.

(c) Reduce engine speed to a slow idle, check the fluid level with the dipstick. With the engine still running, add fluid to bring the level up to the "Full" marking on the dipstick.

Do not overfill; if necessary, drain off to correct level.

(v) Run engine for a few minutes and examine that all seals and joints are oil-tight.

(vi) Replace front floor carpet.

# SERVICE HANDBOOK

### SILVER WRAITH — SILVER DAWN — BENTLEY MK. VI.
### R. TYPE BENTLEY — PHANTOM IV.

## THIRD MOTION SHAFT.

**Strengthened Thrust Washer.**

A thicker thrust washer for the 3rd speed gear is now
fitted on production and is available for fitting when a gearbox is stripped,
if this washer has not already been fitted. Gearboxes already modified can be
identified by the letters "W1" or "W2" stamped on the boss for the second
motion shaft locating screw under the rear end of the box. This washer is
.050" thicker than the earlier type.

**Parts Required.**

RG.5829 Thrust Washer - 3rd Motion Shaft                          1 Off.

RG.8530 3rd Motion Shaft or RG.5469 modified to RG.8530    1 Off.

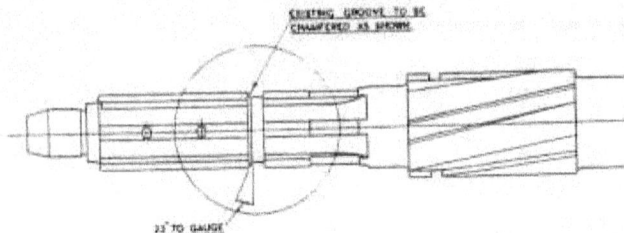

### TO MODIFY THE THIRD MOTION SHAFT
### RG.5469 TO RG.8530.

Grind off the chamfer or the front face of the groove,
as shown in the illustration, using the thrust washer as a gauge. The
washer should rotate freely, but the end float must not exceed .004".

SILVER WRAITH — SILVER DAWN — BENTLEY MK. VI.
R. TYPE BENTLEY — PHANTOM IV.

# SECTION
# G
# BRAKES

## SECTION G.

## B R A K E S

List of Illustrations:

SECTION G.


BRAKES


DESCRIPTION - SETTING OF BRAKE LINKAGES - BRAKE COMPLAINTS -
RELINING BRAKES - OVERHAULING ADJUSTERS - OVERHAULING EXPANDERS -
OVERHAULING MASTER CYLINDER - OVERHAULING SERVO MOTOR.

SECTION G.

B R A K E S

DESCRIPTION:

The four wheel braking system on all models, is of the same type, direct mechanical operation on the rear wheels, and servo-hydraulic operation on the front wheels. The servo motor also operates to increase the braking effort on the rear wheels. Approximately 55 per cent of the total braking effort is imposed on the front wheels, thus overcoming the greater weight thrown on the front wheels during the braking period.

Hand Brake:

The hand brake lever is located under the dash and is connected by a cable to a long idler lever pivoted on the frame, which is in turn connected to the cross-shaft and so to the rear equaliser, to operate the rear brake shoes. The hand brake cable being yoked to one end of the idler lever, provides the proper torque for hand brake application.

Two types of dashboard control levers are fitted, one having a trigger action, see Fig. G1, and the other having a twist release action, see Fig. G2. Referring to Fig. G1, the release trigger is not directly connected to the ratchet pawl, but compresses a spring. This spring in turn, operates the pawl but is only strong enough to move it out of engagement when the load has been removed. Therefore, the hand brake must be pulled ON slightly when releasing the brakes.

On the twist release type, the hand brake is applied by a direct pull, and released by twisting the handle in a clockwise direction.

Figs. G1 and G2, illustrate the layout of the system and the original settings.

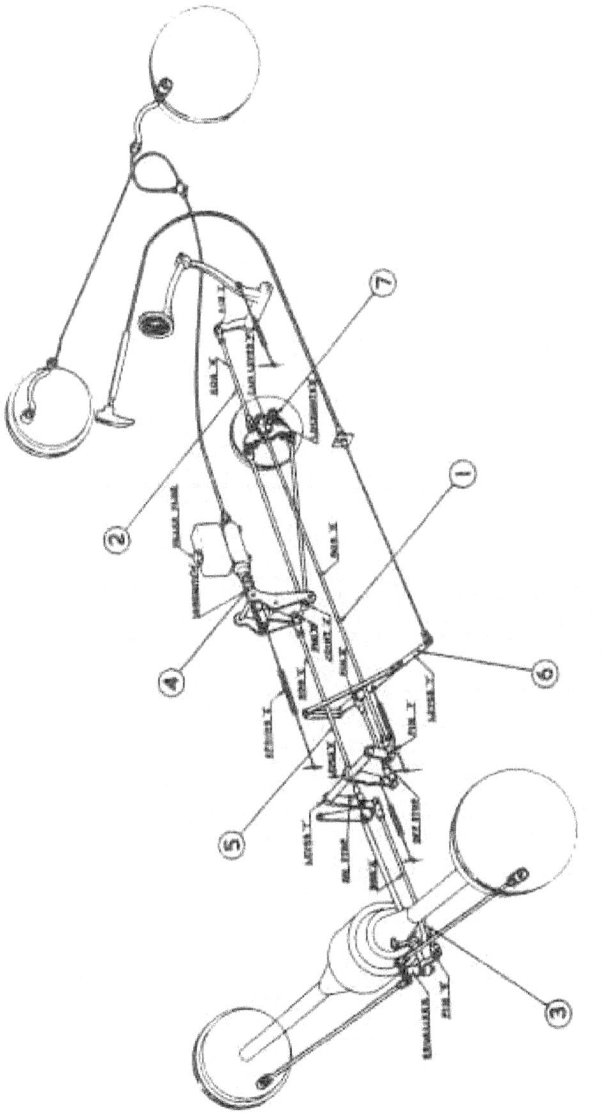

SILVER WRAITH — SILVER DAWN — BENTLEY MK. VI.

R. TYPE BENTLEY — PHANTOM IV.

FIG. G1. BRAKE SETTINGS (RIGHT-HAND CHASSIS)

## SETTING OF BRAKE LINKAGES - RIGHT HAND

The following should be disconnected before starting to set the system. All other components should be assembled.

| | | |
|---|---|---|
| (i) | Pin 'B' | (Rear Equaliser) |
| (ii) | Pin 'D' | (Balance Lever |
| (iii) | Pin 'E' | (Hand Brake |
| (iv) | Pin 'F' | (Inter Shaft Lever) |
| (v) | Pin 'I' | (Pedal Lever |
| (vi) | Spring 'X' | (Master Cylinder ) |

The Setting of Linkages should be carried out in the following sequence:

| | |
|---|---|
| (1) | Adjustment of Rod 'G' (Pull Rod Servo) |
| (2) | Adjustment of Rod 'H' (Pull Rod Servo to Pedal) |
| (3) | Adjustment of Rod 'A' (Pull Rod - Rear Equaliser) |
| (4) | Adjustment of Master Cylinder Push Rod. |
| (5) | Adjustment of Rod 'C' (Pull Rod - Balance Lever) |
| (6) | Adjustment of Hand Brake. |
| (7) | Adjustment of Servo. |

(1)     Adjustment of Rod 'G':

(a)     Move rod 'G' forwards until the limit of swing of jaw on cam lever 'F' has been reached.

(b)     Hold lever 'R' forwards to the limit of its travel and adjust rod 'G' to suit and then shorten rod 'G' by approximately 0.100".

(c)     Replace pin 'F' and lock up nut on rod 'G'.

(2)     Adjustment of Rod 'H':

(a)     This rod can only be adjusted with the pedal gap plate firmly fixed in its normal position.

(b)     Move lever 'R' rearwards on to its off-stop.

(c)     Adjust rod 'H' so that pedal lever, or pedal lever rubber, if one is fitted, is just clear of the spring retainer on gap plate.

(d)     Check this adjustment by pulling upward on the pedal, when rod 'G' should bow slightly before pedal lever, or pedal lever rubber reaches the spring retainer on gap plate.

(e)     Replace pin 'I' and lock up nut on rod 'H'.

(3)     Adjustment of Rod 'A':

(a)     Pull rod 'A' rearward so that lever 'R' is against off-stop and adjust rod 'A' until there is just sufficient tension in rods from rear brakes to equaliser to ensure freedom from rattles.

(b)     Replace pin 'B' and lock up nut on rod 'A'.

(4)  **Adjustment of Master Cylinder Push Rod:**

    (a) Adjust locknuts 'M' so that the clearance between the extending pin 'K' and the master cylinder support bracket 'L' is 1.100" to 1.150" when the plunger contacts the master cylinder piston.

    (b) Tighten locknuts.

(5)  **Adjustment of Rod 'C':**

    (a) Ensure that rear brakes are in off position and lever 'R' is against its off-stop.

    (b) Adjust rod 'C' so that joint 'J' has approximately .000" - .050" free travel before the plunger (of which the inner locknut 'M' is an integral part), contacts the master cylinder piston.

    (c) Replace pin 'D' and lock up nut on rod 'C'.

    (d) Replace master cylinder pull off spring 'X'.

(6)  **Adjustment of Hand Brake:**

    (a) Adjust rod 'N' so that the outer end of horizontal lever 'Y' moves approximately 0.100" to take up the backlash in the intermediate lever 'S'.

(7)  **Adjustment of Servo:**

    (a) Apply the hand brake so that the servo can then be rocked backwards and forwards.

    (b) Adjust locknuts 'O' until friction between the plates can just be felt on rocking the servo.

    (c) Undo the locknuts 'O' 2 to 3 flats until the servo is free again applying the pedal once to ensure that the outer cam lever has followed back the locknuts.

    (d) Tighten the locknuts.

    (e) Check the pedal travel to open the cam levers which should be between 0.250" and 0.500".

SILVER WRAITH — SILVER DAWN — BENTLEY MK. VI.

R. TYPE BENTLEY — PHANTOM IV.

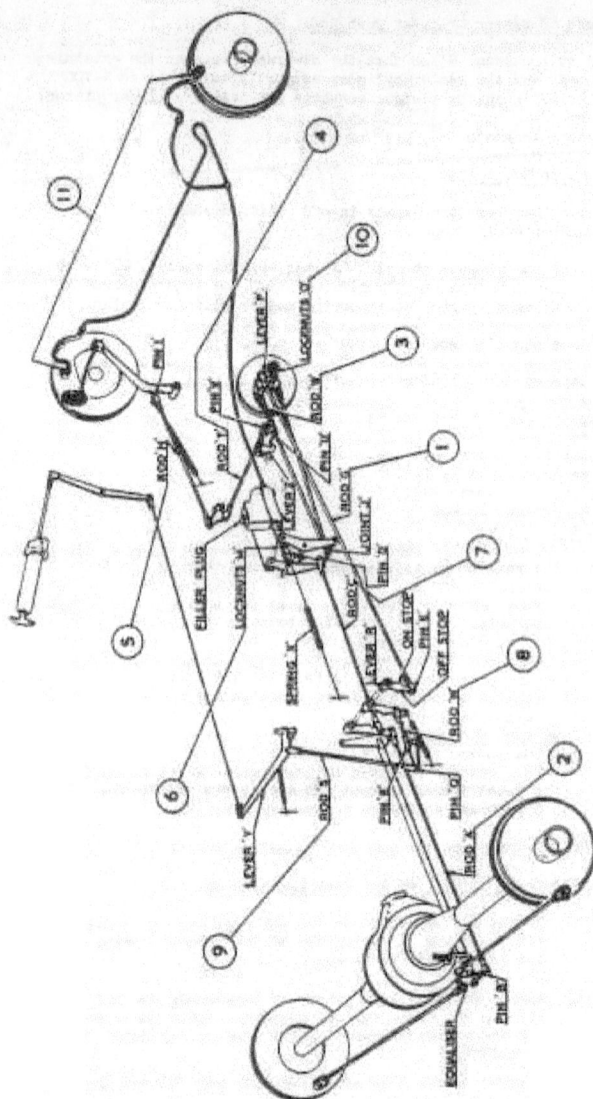

FIG. G2.   BRAKE SETTINGS  (LEFT-HAND CHASSIS)

SILVER WRAITH — SILVER DAWN — BENTLEY MK. VI.

R. TYPE BENTLEY — PHANTOM IV.

SETTING OF BRAKE LINKAGES - LEFT-HAND

The following should be disconnected before starting to set the system. All other components should be assembled.

|       |                                     |
|-------|-------------------------------------|
| (i)   | Pin 'B' (Rod 'A' - Rear Brake Equaliser)  |
| (ii)  | Pin 'D' (Rod 'C' - Balance Lever   )  |
| (iii) | Pin 'E' (Rod 'N' - Bellcrank Lever )  |
| (iv)  | Pin 'Q' (Rod 'S' - Bellcrank Lever )  |
| (v)   | Pin 'F' (Rod 'G' - Lever 'R'       )  |
| (vi)  | Pin 'I'' (Rod 'W' - Lever 'Z'      )  |
| (vii) | Pin 'V' (Rod 'T' - Lever 'Z'       )  |
| (viii)| Pin 'I' (Rod 'H' - Pedal Lever     )  |
| (ix)  | Spring 'X'.                           |

The Setting of the Linkage should be carried out in the following sequence:

|       |                                                          |
|-------|----------------------------------------------------------|
| (1)   | Adjustment of Rod 'G' (Servo - Lever 'R'    )            |
| (2)   | Adjustment of Rod 'A' (Rear Brake Equaliser)            |
| (3)   | Adjustment of Rod 'W' and Stop Lamp Switch.              |
| (4)   | Adjustment of Rod 'T' (Bellcrank Lever - Lever 'Z'  )   |
| (5)   | Adjustment of Rod 'H' (Pedal Lever - Bellcrank Lever)   |
| (6)   | Adjustment of Master Cylinder Push Rod.                  |
| (7)   | Adjustment of Rod 'C' (Balance Lever - Intershaft Lever  ) |
| (8)   | Adjustment of Rod 'N' (Intershaft Lever - Bellcrank Lever) |
| (9)   | Adjustment of Hand Brake.                                |
| (10)  | Adjustment of Servo:                                     |

(1)   Adjustment of Rod 'G':

   (a)  Move rod 'G' forward until the limit of swing of the jaw on cam lever 'P' has been reached.

   (b)  Hold lever 'R' forward against its on-stop, and adjust rod 'G' to suit.

   (c)  Make rod 'G' approximately 0.100" shorter than this.

   (d)  Replace pin 'F' and lock up nut on rod 'G'.

(2)   Adjustment of Rod 'A':

   (a)  Pull rod 'A' rearward so that lever 'R' is against its off-stop, and adjust rod 'A' until there is just sufficient tension in rear brake rods to ensure freedom from rattles.

   (b)  Replace pin 'B' and lock up nut on rod 'A'.

(3)   Adjustment of Rod 'W' and Stop Lamp Switch:

   (a)  Locate the lever 'Z' in its off position by lining up with the hole provided in the bottom of the bracket, using the pin 'V' and adjust rod 'W' to suit.

   (b)  Adjust the stop lamp switch by slackening the two 2 BA bolts and sliding the whole unit as required, until the clearance between the end of the plunger and the boss on the lever 'Z' is 0.020" to 0.030".

   (c)  Tighten up the 2 BA bolts, replace pin 'U' and lock up nut on rod 'W'.

(4)  Adjustment of Rod 'T':

    (a)  Adjust rod 'T' to the nearest half turn of the jaw to give 12.250" centre distance between the two pins.

    (b)  Replace pin 'V' and lock up nut on rod 'T'.

(5)  Adjustment of Rod 'H':

    (a)  This rod can only be adjusted with the pedal gap plate firmly fixed in its normal position, and therefore must be finally set with the body fitted.

    (b)  Ensure that lever 'R' is on its off-stop.

    (c)  Adjust rod 'H' so that the pedal lever is just clear of the gas seal on the gap plate.

    (d)  Check this adjustment by pulling upward on the pedal when rod 'G' should bow slightly before the pedal lever reaches the gap plate.

    (e)  Replace pin 'I' and lock up nut on rod 'H'.

(6)  Adjustment of Master Cylinder Push Rod:

    (a)  Adjust locknuts 'M' so that the clearance between the extended pin 'K' and the master cylinder support bracket 'L' is 1.100" to 1.150" when the push rod contacts the master cylinder piston.

    (b)  Tighten the locknuts.

(7)  Adjustment of Rod 'C':

    (a)  Ensure that rear brakes are in off position, with lever 'R' on its off-stop.

    (b)  Adjust rod 'C' so that joint 'J' has 0.000" to 0.050" free travel before the master cylinder push rod contacts the piston.

    (c)  Replace the pin 'D' and lock up nut on rod 'C'.

    (d)  Replace master cylinder pull off spring 'X'.

(8)  Adjustment of Rod 'N':

    (a)  Adjust rod 'N' to the nearest half turn of the jaw to give 5.375" centre distance between the two pins.

    (b)  Replace pin 'E' and lock up nuts on rod 'N'.

(9)  Adjustment of Hand Brake:

    (a)  Adjust rod 'S' so that the outer end of the horizontal lever 'Y' moves approximately 0.100" to take up the backlash and begins to move rod 'A'.

    (b)  Replace pin 'Q' and lock up nuts on rod 'S'.

(10)   Adjustment of Servo:

  (a)  Apply the hand brake, so that the servo can be rocked backwards and forwards.

  (b)  Adjust locknuts 'O', until friction between the plates can just be felt on rocking the servo.

  (c)  Undo the locknuts 'O', 1 to 2 flats until the servo is free again, applying the pedal once to ensure that the outer cam lever has followed back the locknuts.

  (d)  Tighten the locknuts, and check that the servo is still free.

## SERVICING:

It should be appreciated that a thorough knowledge of the system, absolute cleanliness and careful workmanship are necessary. Any foreign matter in the hydraulic system will tend to clog the lines, ruin the rubber cups in the wheel and master cylinders, and cause inefficient operation of the front brakes.

Likewise dirt or grease on a brake lining will cause that lining to take effect on brake application and fade out on heavy brake application.

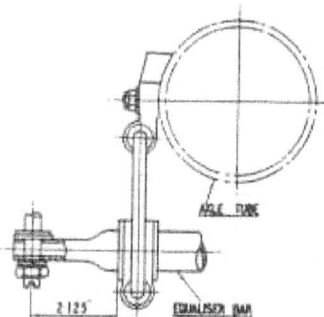

FIG. G3.   CORRECT POSITION OF REAR EQUALISER SUPPORT.

The need for adjustment of the rear brakes will be indicated by excessive travel of the hand brake lever. The front brakes, operated by the servo motor only, will not affect the hand brake or pedal, but providing these are adjusted at the same time as the rear brakes, no trouble should be expected.

Low or inconsistent output from the servo would be indicated by heavy or non-progressive brake pedal action, together with insufficient front braking. The remedy is to dismantle the servo to find the trouble, see page G23.

A light pedal action, accompanied by defective front braking with rear wheels locking, would indicate fault in hydraulic system. Excessive front braking would indicate fault in rear brakes.

(1)   Adjustment of Brakes:

A separate adjuster is provided on each brake carrier plate and these are the only adjustments provided on the whole system. No alteration must be made to brake rods. See Figs. G4 and G5.

The method of adjustment is the same for both front and rear brakes, except that a hand wheel is fitted for rear brake adjustment and a $\frac{1}{4}$" B.S.F. spanner working on a square headed adjuster is necessary for the front brakes.

For each complete turn of the adjuster screw, four "clicks" will be felt, and between each "click" the brake shoes are expanded approximately, .014" and then moved back .010" to give an adjustment of .004" and a running clearance of .010".

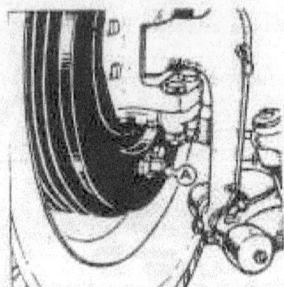

FIG. G4.  HAND ADJUSTMENT (FRONT BRAKES)
A. Adjuster Screw.

FIG. G5.  HAND ADJUSTMENT (REAR BRAKES)
B. Adjuster Screw.

Rotate the adjuster screw in a clockwise direction until considerable resistance is felt.  The resistance must be equal for all four brakes, should any one adjuster require noticeably greater force to obtain the last "click", turn back to previous "click".

Adjustment of the rear brakes takes up both the pedal and hand brake clearance in the same operation.

If the adjuster screws come to the end of their travel, indicated by a solid resistance feel, the brake linings will need renewal.

The hand brake lever should be adjusted so that it has a pull of five notches to engage brakes, adjust on Rod 'N'.

The servo motor is of the disc brake type.  An adjustment is provided for initial setting and as very little wear occurs, no normal servicing is necessary.  For overhauling, see page G23.

(2)  Bleeding Hydraulic System:

Bleeding, i.e., expelling air from the system, should only be necessary when completely recharging the system with fluid after the removal of a component, or the disconnection of a pipe joint.

Under normal conditions, air does not enter the system as a result of brake application.

Carefully clean all dirt from around the filler plug on the master cylinder.

Remove filler plug, install automatic filler or fill up master cylinder using new oil, Fig. G6.

Attach drain hose to bleeder valve, and release valve screw half to one turn.

Keep end of drain hose immersed in fluid jar.  Fig. G7.

Push the joint J, Fig. G6, forward until the extended pin K, abuts against the master cylinder support brackets.

Release, pause and repeat until all bubbles cease to appear at end of drain tube, take care not to entirely empty master cylinder.

Tighten bleeder screw and transfer drain tube to the other front brake and repeat.

Finally, refill reservoir and replace plug.

NOTE: Only Lockheed Brake Fluid - Orange - should be used - A mixture containing mineral oil must never be used, as this will rapidly destroy the sealing qualities of the rubber cups.

In the event that improper fluid has inadvertently been used, it will be necessary to:-

(i)    Drain the entire system.

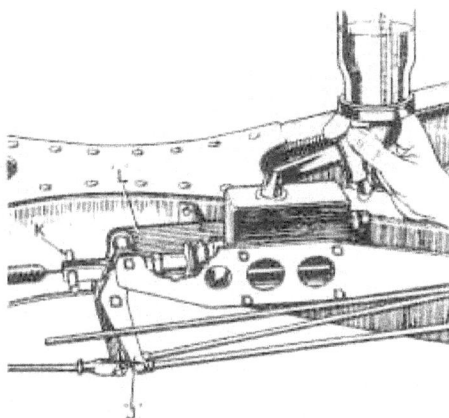

FIG. G6.   FILLING MASTER CYLINDER.

(ii)   Thoroughly and vigorously flush the system with clean alcohol.

(iii)  Replace all rubber parts of the system.

(iv)   Refill with correct fluid.

BRAKE COMPLAINTS:

In an effort to overcome braking complaints on Bentley Mk.VI and Silver Wraith cars, certain modifications to the linkages and wedge angles have been introduced. See Fig. G8.

Referring to Fig. G8, pedal effort and front/rear brake ratio can be varied by changes in the servo cam angle 'A', the length of the balance lever 'B', and the expander wedge angle 'C'. Original cars were fitted with servo cams machined to $30^\circ$ (angle 'A'), a balance lever in which dimension 'B' was 3" and expander wedge angles of $9^\circ$ at 'C'. The resulting front ratio was 1.37. To reduce front braking, dimension 'B' of the balance lever was altered to 2.4", which gives a front/rear ratio of 1.17. The next alteration was made to minimise the possibility of rear brakes sticking after release; this consisted of changing the expander wedge

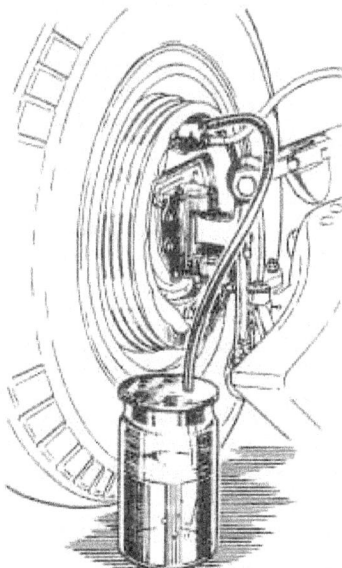

FIG. G7.   BLEEDING FRONT BRAKE.

## SERVICE HANDBOOK

SILVER WRAITH — SILVER DAWN — BENTLEY MK. VI.
R. TYPE BENTLEY — PHANTOM IV.

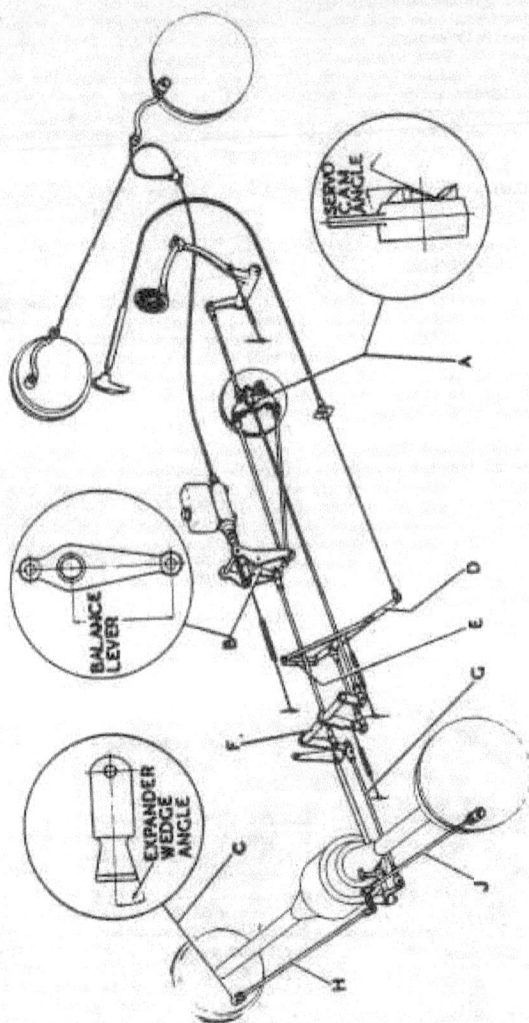

FIG. 98. BALANCE LEVER AND WEDGE ANGLES.

# SERVICE HANDBOOK

### SILVER WRAITH — SILVER DAWN — BENTLEY MK. VI.
### R. TYPE BENTLEY — PHANTOM IV.

angle 'C', from $9^\circ$ to $13^\circ$. This increased the pedal effort required, and it was found necessary to modify the servo cam angle 'A' to $25^\circ$ to restore original operation. At the same time, alterations were made to the hand brake leverage for similar reasons. Later in the same chassis series, the three pull rods connected to the rear brake equaliser were modified to obtain an improved operating arc, and subsequently wedge return springs were included in the rear brake expander units. The proportion of rear braking on cars fitted with $25^\circ$ servo cams, and 2.4" balance levers was considered slightly too high, later series therefore and current production is with 3" balance lever with $22\frac{1}{2}^\circ$ servo cams and $13^\circ$ expander wedges. (NOTE: The angular degree is etched on each appropriate part.)

Silver Dawn chassis are all fitted with a 3" balance lever, with $22\frac{1}{2}^\circ$ servo cams and $13^\circ$ expander wedges.

Phantom IV, chassis are all fitted with 2.4" balance lever, with $20^\circ$ servo cams and $9^\circ$ expander wedges.

Commencing on Bentley MK.VI chassis B-1-GT, Silver Wraith chassis WME-1, Silver Dawn chassis SCA-1, and all Phantom IV chassis, a new type of front brake is fitted, Fig. G11. This differs chiefly in the use of an external cylinder (expander) operating the shoes through a wedge mechanism. Unlike the previous brake, the cylinder is rigidly attached to the carrier plate by three studs. Although the cylinder body is fixed, the loads on the two shoe tips are balanced by virtue of the column of oil between the two pistons.

The equal wear linkage between the two shoes also differs from the previous arrangement, the shorter of the two links is eliminated, one arm of the balance lever being attached directly to the web of the trailing shoe while a long link connects the other arm to the web of the leading shoe. The balance lever, instead of having a fulcrum consisting of a pin attached to the carrier plate, is pivoted on a short push rod which abutts the cylinder body, so that the whole assembly of shoes and linkage is mounted off the cylinder. The bedding of the two shoes can thus be equalised without any adjustment of the inter-shoe linkage being necessary.

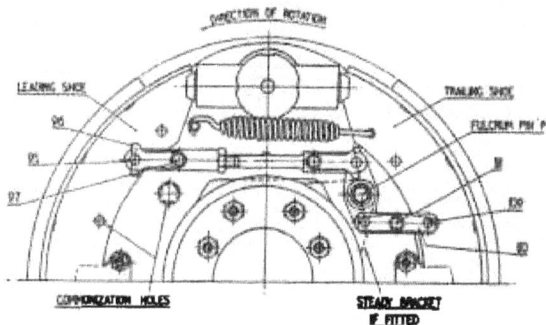

FIG. G9. EQUAL WEAR LINKAGE (ORIGINAL)

SILVER WRAITH — SILVER DAWN — BENTLEY MK. VI.
R. TYPE BENTLEY — PHANTOM IV.

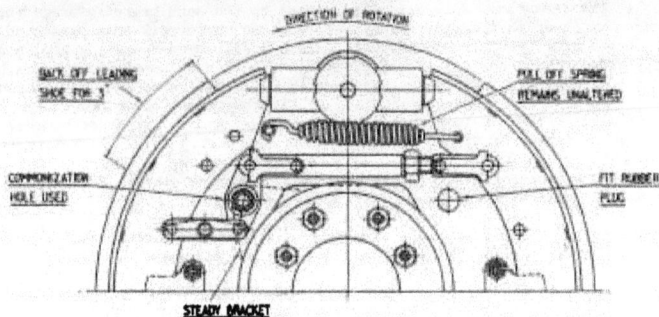

FIG. G10. EQUAL WEAR LINKAGE (REVERSED LINKAGE).

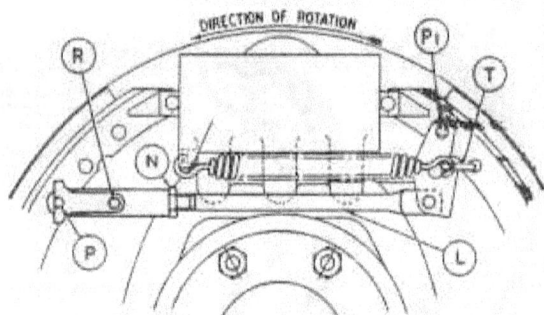

FIG. G11. EQUAL WEAR LINKAGE (PRESENT PRODUCTION).

The usual complaints are:-

(i)     Brakes fierce on first application,
        Too much front braking,
        "Spragging" (locking at low speeds).

(ii)    Inefficient front brakes.

(iii)   "On" clonks or "Off" clonks.

(iv)    Uneven or "pulling to one side".

(v)     Front brake rumble.

# SERVICE HANDBOOK

## SILVER WRAITH — SILVER DAWN — BENTLEY MK. VI.
## R. TYPE BENTLEY — PHANTOM IV.

Remedy:-

Before any adjustment is made, check that the tyre pressures are correct, and examine tread on tyres for any excessive wear on any one wheel.

(i)  Remove drums and check linings, back off leading and trailing shoes to second rivet. On early type fit later type servo cam and 2.4" balance lever.

Check inter-shoe linkage, on early type this may be due to trailing shoe pivoting about the inter-shoe linkage instead of the expander plunger. Fit reverse type linkage, see Fig. G10.

(ii) Check rubber piston cup in master cylinder, examination may find that this is grooved and so allows the oil to leak back. Replace.

(iii) See Fig. G1. - Check that rear brakes are in the "Off" position, lever on end of Rod 'G' is against "Off" stop.

Remove master cylinder pull-off spring 'X' and clevis pin 'D'.

Adjust Rod 'C' so that joint 'J' has zero to .050" free travel before plunger 'M' contacts master cylinder.

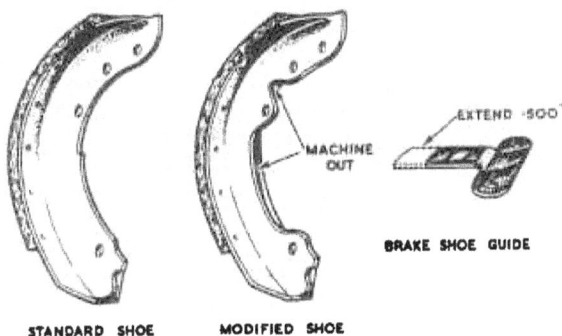

STANDARD SHOE        MODIFIED SHOE

FIG. G12. MODIFICATION TO BRAKE SHOE.

(iv) Check drums for concentricity, reverse shoes and back off, if necessary. Check expander cones for ridging and replace if necessary.

(v)  This has been shown to be due to temporary brake drum distortion due to differential expansion after a series of heavy applications of the brakes at high speeds. The remedy is to check the equal wear linkage and to fit modified brake shoes, if necessary, see Fig. G12, and also longer guides.

## RELINING BRAKES:

Sets of pre-finished brake shoes on which new linings have been fitted and ground to size are available, and should always be fitted when replacement of brake linings are necessary.

NOTE: On cars that have been fitted with the modified brake shoes, it is advisable that the new shoes should be similarly modified before fitting.

## To Dismantle:

(i)    Remove hub caps and wheel discs, slacken off wheel nuts, jack up and remove wheels.

(ii)    Release hand brake and slacken off brake adjusters - anti-clockwise - remove the three counterheaded setscrews from face of drum. Remove hub extension from rear hubs.

(iii)    Withdraw the drum, use two $\frac{1}{4}$" x 26 T.P.I. setscrews as extractors in the holes provided.

(iv)    Where steady bracket is fitted to grease catcher assembly, withdraw front hub by removing split pin and nut on stub axle, and remove grease catcher assembly by undoing the eight nuts.

## Front Brakes:

See Figs. G9, G10 and G11.

To remove the brake shoes, disconnect the equal wear linkage, this may be as Fig. G9 or G10, according to whether the linkage has been reversed to cure a complaint of "spragging".

(i)    Remove setscrews (111), locking plate (110) and by unhooking pull-off spring, withdraw pin (109).

(ii)    Remove setscrews (97) locking plate (96) and withdraw pin (95). Remove equal wear linkage complete from fulcrum pin.

(iii)    Prise brake shoes from expander slots and remove, disconnect lower pull-off spring.

As previously mentioned, after Bentley MK.VI chassis B-1-GT, Silver Wraith chassis WME-1, Silver Dawn chassis SCA-1, and all Phantom IV chassis, new types of brakes were introduced, see Fig. G11.

To remove the brake shoes:-

(i)    Unhook the pull-off spring, taking care not to lose the small hook (T). Remove setscrew (R) and the locking plate.

(ii)    Remove the pins (P) and (P1) and then remove the linkage. The shoes may then be prised out from the expander lugs and removed.

## Rear Brakes:

To remove the rear brake shoes, proceed as described for the front brakes.

## Re-assembling: (Early type)

NOTE: Keep all grease off linings, do not handle more than necessary. Inspect pull-off spring and replace if stretched or damaged.

SILVER WRAITH — SILVER DAWN — BENTLEY MK. VI.

R. TYPE BENTLEY — PHANTOM IV.

(i)     Detach pull-off spring from old shoes, and fit lower spring to new
        shoes. Fit the other spring to the trailing shoe and anchor other
        end to pillar.

(ii)    Fit the trailing shoe first into the slot of the expander plunger and
        the other end into the adjuster unit. Repeat for leading shoe.

(iii)   Replace the equal wear linkage, minus locking plate.

Adjustment of Linkage:

NOTE:   The front hydraulic
        expanders should be
        free to slide to and
        fro on the elongated
        holes in the carrier
        plate.

(i)     From a piece of light
        gauge steel, make up
        a trammel as shown in
        Fig. G13, to fit over
        two of the wheel studs
        on the hub, and bend
        the pointer to touch
        lightly on the face of
        the brake linings.

(ii)    Temporarily replace the
        brake drum, operate the
        brakes and externally
        adjust the brake shoes.

FIG. G13.   TRAMMEL FOR BRAKE SHOE ALIGNMENT.

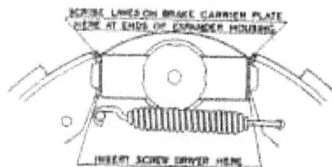

(iii)   Remove brake drums and fit trammel on any two wheel studs.

(iv)    By making suitable adjustments to the brake equaliser push rod,
        centralise the brake shoes, rotating the trammel about the shoes
        as a gauge.

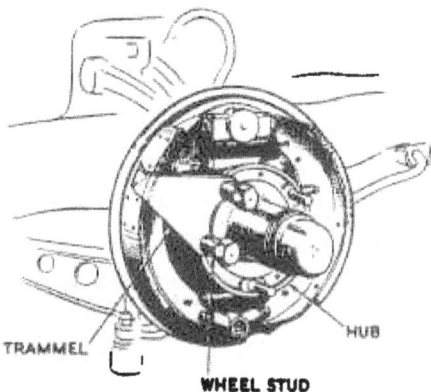

FIG. G14.   CHECKING FLOAT - EXPANDER UNIT.

Having corrected bedding of brake
shoes, set angular position of adjuster.
The retaining setscrews pass through
clearance holes in carrier plate, allowing
adjuster to rotate slightly about its
spigot.

(i)     Tighten adjuster to lock brake
        shoes on drum. Tighten retaining
        setscrews, then slacken back
        adjuster screw one notch at a time
        until brake drum is free.

As previously mentioned the expander unit is free to slide. There should
be a minimum float of .060" to either side. Check as follows:-

(i)     Remove brake drum and scribe lines on carrier plate, see Fig. G14.
        Lever with screwdriver to free shoe tip and note free travel.

- G.16 -

SILVER WRAITH — SILVER DAWN — BENTLEY MK. VI.

R. TYPE BENTLEY — PHANTOM IV.

Final Assembly:

Before final assembly, check brake shoe steady brackets. The faces of the brackets contacting the brake shoe webs should be in same plane, i.e., in line with the side faces of the shoe slots in the expander and adjuster plungers, on the side of the slots nearest to the carrier plate. It is permissible for the faces to be .015" below the plane (nearer to the carrier plate), they must never be above it (away from the carrier plate).

(i)     Check over fitting and tighten setscrews.

(ii)    Rear brakes - Replace water excluder and secure. Roll rubber washer into groove and replace rubber dust cover, connect up rod and secure. Correct adjustment of brakes.

Re-assembling:  (Later type)

The same remarks on absolute cleanliness of the linings, as previously stated, apply.

The brake should be built up in a similar manner as for the previous model. Make sure that the spring hook 'T' Fig. G11, is correctly in position.

Adjustment of Linkage:

(i)     Build up brake complete, except for locking plate, pin (P). For initial setting, link (L) should be set to 7.375" between centres.

(ii)    Fully screw back hand adjuster and slacken setscrews securing adjuster to carrier plate. Fit brake drum; if drum cannot be fitted over shoes, lengthen link (L) by detaching pull-off spring and withdrawing pin (P).

(iii)   With drum in position, check total number of "clicks" obtained on adjuster to bring shoes in contact with drum, which should be between 9 and 15. If this is not so, adjust link (L) to suit. (One complete turn of jaw, approximately three "clicks" of adjuster.)

(iv)    Next, centralise assembly by screwing up adjuster until shoes are hard on drum, then tighten up nuts securing cylinder. Slacken back adjuster, remove drum, fit locking plate and look up. Replace drum.

NOTE: After relining, to allow for "growth" of the new linings due to temperature effects, the external adjustment should be set back six or seven "clicks" for the initial road test. After road test, reset adjusters and finally turn back three "clicks". This is at variance with the driver's Handbook, but the linings will be stabilised by the time the need for adjustment arises.

OVERHAULING ADJUSTERS:

The adjusters for all four brakes are of similar internal construction, and it is recommended that when the brakes are relined that these should be removed, dismantled, cleaned and greased.

(i)     Remove the adjuster and dismantle by withdrawing the plungers and screwing the adjuster screw inwards. Clean and grease.

(ii)    To re-assemble, replace adjuster screw in housing, then correctly replace the plungers. The plungers are handed and, correctly assembled, the inclined faces on the inner ends must be parallel along the axis of the adjuster screw. With plungers correctly assembled, there are four evenly spaced "clicks" per turn of the adjuster screw.

(iii)   If rubber seal is in poor condition, renew. Check clearance between setscrew and plate on plungers.

OVERHAULING EXPANDERS:

Front Brake: (Early type) Fig. G16.

The only part that may require replacing, is the rubber cup (72).

To dismantle, remove pins (66) which will allow removal of plungers (68) and roller (60). Remove grub screw (61) and with 'C' spanner, remove threaded cap (62) with spring (63). Remove expander core (65), rubber cup (72) and spring (74) with end cap (73). Remove bleeder screw, retain steel ball (valve).

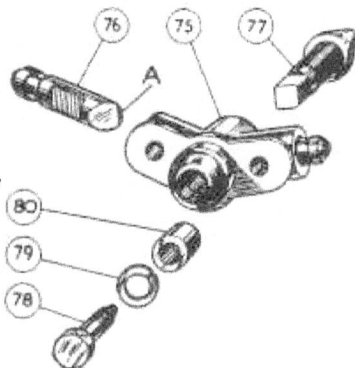

FIG. G15.   FRONT ADJUSTER.

75. Plunger Housing
76. Plunger
77. Adjuster Screw
78. Setscrew
79. Spring Washer
80. Distance Piece

It is essential that no mineral oil or grease is allowed to contact the internal parts.

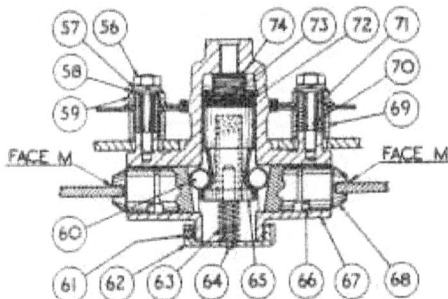

FIG. G16.   FRONT EXPANDER (Early type).

To re-assemble, replace spring (74) complete with end cap. Enter rubber cup (72), with lip inwards, into cylinder about ½" followed by expander cone, hole outermost. Guide lip of rubber cup past intersection by inserting finger in plunger bore, hold expander cone in position and re-assemble the two rollers and plungers.

Replace the spring (63) in the expander cone, replace the cap, tighten up and lock with grub screw. Fit hose assembly, replace steel ball and bleeder screw.

Refit into position in carrier plate, using new seal if necessary.

Front Brake: (Later type) Fig. G17.

The only parts that may need replacing are the rubber seals, cups or spring. Unless there are evident signs of a leak, the expander should not be dismantled.

To dismantle - Extract the rubber seals from the ends, the pistons, rubber cups and centre balance spring can then be pushed out.

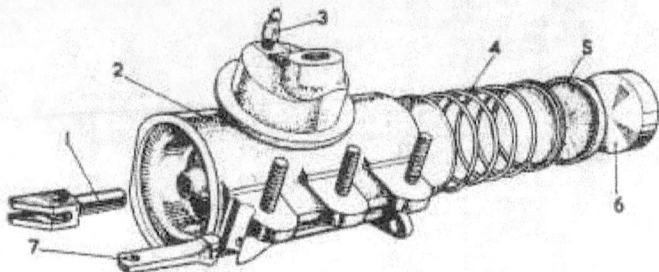

FIG. G17. FRONT EXPANDER. (Later type)

| | |
|---|---|
| 1. Operating Link | 5. Rubber Cup |
| 2. Housing, Cylinder | 6. Piston |
| 3. Bleeder Connection | 7. Pivot Link |
| 4. Spring | |

To re-assemble - Reverse the above instructions, refer to Fig. G17, for correct assembly.

## OVERHAULING REAR EXPANDERS:

It is recommended that the rear expanders are removed, dismantled, cleaned and greased when the brakes are relined.

To remove - Release the brake rod from the equaliser lever, remove dust cover and rubber ring from expander housing.

Remove hand wheel from adjuster, unscrew four nuts retaining water excluder and remove. Remove expander retaining nuts and withdraw.

Remove plunger retaining pins and withdraw with rollers.

Remove dust cap; push brake rod with the expander cone through the housing, tap out retaining pin (10).

To re-assemble - Lubricate all internal parts with hub grease.

Re-assemble by reversing the removal instructions. Secure the dust cap by swaging the metal in three places.

## OVERHAULING MASTER CYLINDER:

To remove from chassis - Disconnect the operating rod (44) Fig. G20. Remove nut, bolt and distance piece from joint 'J', thus disconnecting servo pull rods. Disconnect pull-off spring and hydraulic pipe. Remove retaining bolts and remove mounting bracket complete with unit from chassis.

SILVER WRAITH — SILVER DAWN — BENTLEY MK. VI.

R. TYPE BENTLEY — PHANTOM IV.

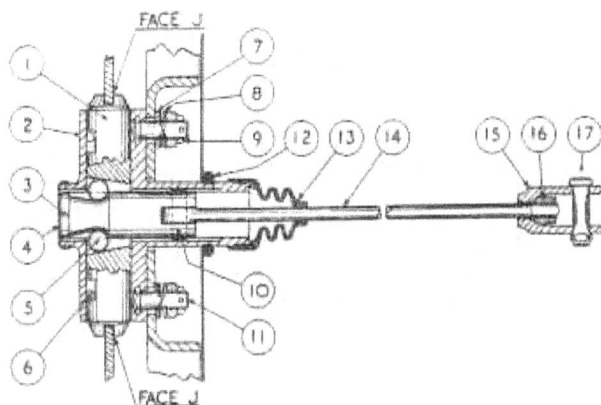

FIG. G18. REAR EXPANDER.

| | |
|---|---|
| 1. Plunger | 6. Retaining Pin - Plunger |
| 3. Expander Cone | 10. Pin - Cone |
| 4. Dust Cap | 14. Brake Rod |
| 5. Roller | |

To dismantle:-

(i)   Remove filler plug and drain.

(ii)  Remove rubber boot, extract piston stop retaining ring from groove, then remove piston stop washer. Remove remaining components for inspection.

Servicing - Replace any rubber components that are not perfect. If either of the valve rubbers, (12 and 13) Fig. G19. need replacement, replace complete valve assembly.

Thoroughly clean, and make sure that the two holes in the base of the reservoir are clear.

1. Push Rod
2. Boot
5. Piston
6. Secondary Cup
8. Main Cup
9. Retainer for Spring
10. Spring - Piston Retainer
11. Valve Assembly
14. Filler Plug

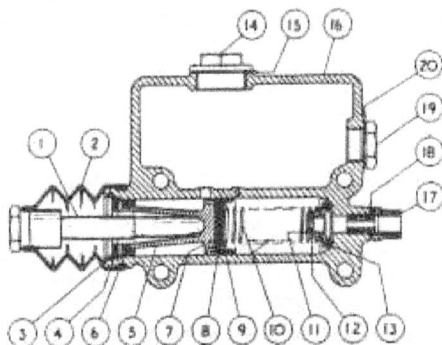

FIG. G19. MASTER CYLINDER.

- G.20 -

To re-assemble:-

(i)     Mount unit in vice with discharge end downwards.

(ii)    With valve assembly attached to spring, lower into cylinder, check that
        end is central.

(iii)   Insert rubber cup (8) and press firmly onto retainer.  Insert piston
        washer (7).

(iv)    With secondary cup (6) assembled as shown in Fig. G19, with cam, insert
        in cylinder.  Press piston down, replace retaining washer and retaining
        ring.  Refit rubber boot.

        To refit the master cylinder, reverse the removal instructions.

## OVERHAULING SERVO MOTOR:

Removal from chassis:-

(i)     Remove undersheet.

(ii)    Refer to Fig. G20.  Remove pin (39) and unscrew rod 'H' from jaw (37).
        Remove pin (52) and unscrew rod 'G' from jaw (36).  Remove bolt (48)
        and free drag links.  Rotate servo to uncouple front drag links.

(iii)   Remove setscrews, locking plates and pins (41, 43 and 40).  Remove servo
        retaining screw (31, Fig. G21) and carefully pull-off servo from driving
        pins.

(iv)    Next, manoeuvre servo with outer end turned sideways to the front of car
        and inside cruciform, to draw clear.  Remove Ferodo washer from gearbox.

        NOTE: If there is not sufficient clearance for withdrawal, the rear end
              of the gearbox will have to be moved sideways for approximately
              half an inch.

              Slack off inner nut on torque reaction arm on L.H. side and remove
              rubber.  Remove the two bolts from the rear end cover and support
              bracket.  With the gear lever in Neutral, lever gearbox about ½"
              to L.H. side of chassis.

Servicing:-

Possible causes of trouble:-

(i)     Oil or foreign matter on friction faces.

(ii)    Dirt or lack of oil on working parts.

(iii)   Damage to shaft on outer race of ball bearings.

(iv)    Indentation of cam faces by balls.

(v)     Excessive clearance due to wear of friction face or incorrect adjustment.

        If oil is found on the friction faces it is essential to find the cause
and rectify.  Check gearbox oil seal.

SILVER WRAITH — SILVER DAWN — BENTLEY MK. VI.

R. TYPE BENTLEY — PHANTOM IV.

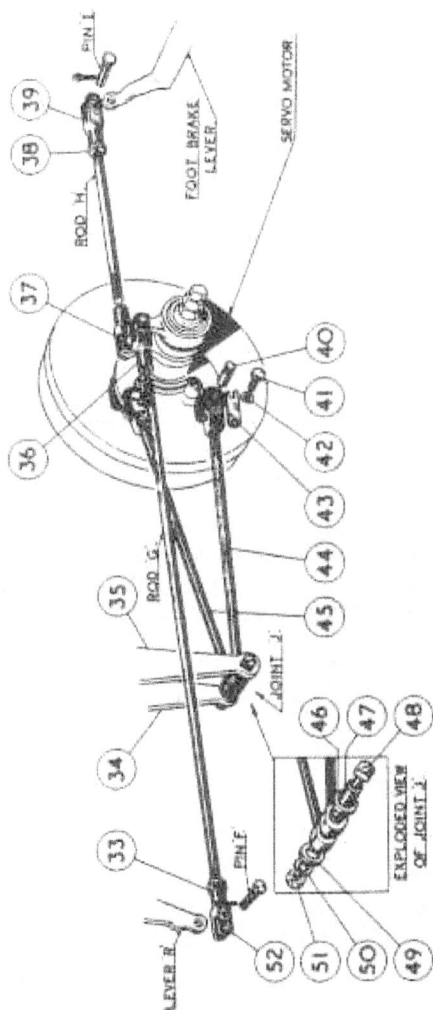

FIG. G20. REMOVING SERVO MOTOR.

Dismantling servo:-

(i) Pull off protector (20, Fig. G21), this will allow driving plate assembly, (9, 10, 15 and 17) and spring plate (12) to be separated from rest of servo.

> NOTE: No further dismantling is usually necessary. Check "Bellville" washer retaining inertia ring has not sprung back and so relieving loading; press around circumference to snap back into position.

(ii) Mount shaft in vice and remove the two nuts (32 and 30). Remove outer lever complete with ball bearing, collect the three balls, .3125" dia.

(iii) Remove inner lever (26), ball cage, dust cover and thrust bearing assembly (25 and 3). Remove lever (23) and lever (6).

(iv) Remove pressure plate, invert to remove ball cage. Examine and replace any damaged parts.

Relining servo:-

Rotate spring damper inertia ring to line up access holes with rivet heads, drill out with 5/32" dia. drill and tap out using ⅛" pin pinch.

Remember "Ferodo" is brittle, re-rivet with care. Clean up face by rubbing down on coarse glass paper.

Re-assembling servo:-

Re-assemble by reversing dismantling procedure.

Check:-

(i) Inner race of ball bearing fitted to outer lever is sliding fit on servo shaft.

(ii) Servo shaft is sliding fit in spigot.

Refitting servo to gearbox:-

Reverse the removal procedure:-

Check:-

(i) Ferodo sealing washer on gearbox has the VG.100 (black coloured) face towards servo.

(ii) Flange of servo shaft has fully engaged with driving pins.

To adjust servo:-

(i) Apply hand brake.

(ii) Adjust nut (30, Fig. G21) until friction can be felt, between pressure plate and lining. While rocking servo, slacken back ½ turn, giving clearance of approximately .018" 'A' Fig. G21. Apply pedal once to ensure outer cam lever has followed back.

(iii) Hold setting and lock (32). Check pedal travel to open cam levers, .250" to .500".

Servo Seal 1 3/4" shaft, 2 1/2" outside diameter.

SILVER WRAITH — SILVER DAWN — BENTLEY MK. VI.

R. TYPE BENTLEY — PHANTOM IV.

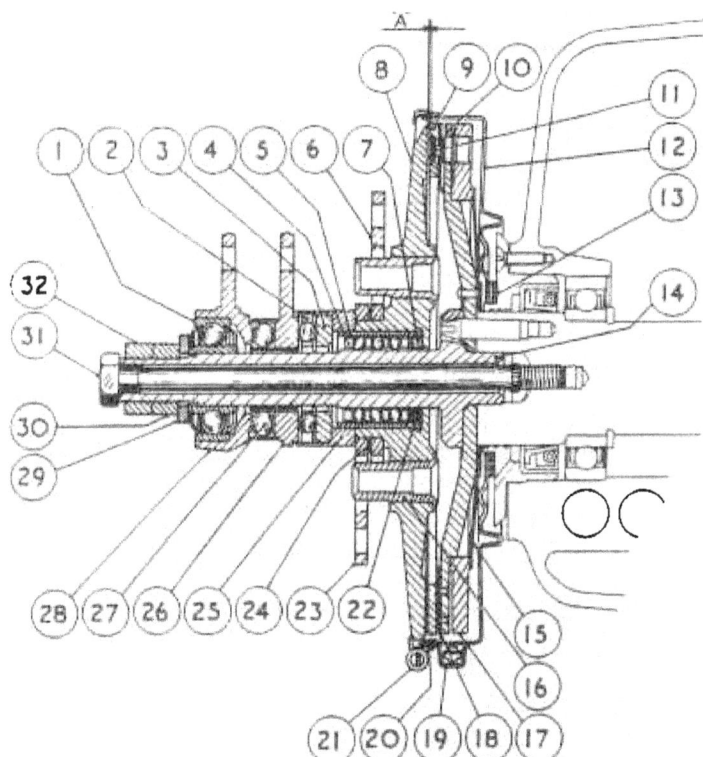

FIG. G21. SERVO MOTOR.

| | | |
|---|---|---|
| 1. | Ball Bearing - outer lever. | |
| 2. | Cage - thrust bearing - cam levers | }Assy: |
| 3. | Thrust race - cam levers | |
| 4. | Cage - ball bearing. | |
| 5. | Ball race - pressure plate. | |
| 6. | Inner lever - brake actuating. | |
| 7. | Felt washer, oil retaining. | |
| 8. | Pressure plate. | |
| 9. | Friction lining. | |
| 10. | Driving plate. | |
| 11. | Rivet - plate to lining (9 off). | |
| 12. | Spring plate. | |
| 13. | Sealing washer - spring plate. | |
| 14. | Servo shaft. | |
| 15. | Bellville washer - servo. | |
| 16. | Operating pin - pressure plate. | |
| 17. | Inertia ring - servo. | |
| 18. | Spring plate (9 off). | |
| 19. | Water drain - servo (assy). | |
| 20. | Protector - servo. | |
| 21. | Spring - protector. | |
| 22. | Washer - felt retaining. | |
| 23. | Outer lever - brake actuating. | |
| 24. | Washer - brake actuating lever. | |
| 25. | Dust cover - thrust bearing. | |
| 26. | Inner lever - servo operating. | |
| 27. | Ball .3125" dia. (3 off). | |
| 28. | Outer lever - servo actuating. | |
| 29. | Plain washer - servo shaft. | |
| 30. | Adjusting nut - servo. | |
| 31. | Setscrew - servo retaining. | |
| 32. | Locking nut - servo adjusting. | |

SILVER WRAITH — SILVER DAWN — BENTLEY MK. VI.

R. TYPE BENTLEY — PHANTOM IV.

FRONT BRAKE EXPANDERS AND MASTER CYLINDER.

Front Expanders.

Instances have occurred of troubles which have been due to leakage of brake fluid from the internal front brake expanders. The bores of these expanders were found to be scored or corroded, while some rubber cups were found to be deficient of radial pressure due to being a slack fit in the bore permitting air to enter the system, causing servo judder, inefficient brakes or brakes pulling to one side. Excessive loss of brake fluid can render the brakes inoperative.

If on removal of the drums, there are signs of leakage, the expander must be removed.

(1)    If the expander is scored and/or corrosion pitted, a new unit must be fitted. Where the bore has only slight scratching it can be polished with Dunex abrasive paper No.240, but it is essential that the bore has a smooth finish.

(2)    Fit new rubber cups (which should be a light push fit) and renew the rubber dust covers if necessary.

(3)    Cases have occurred of air entering the system due to the non-sealing of rubber valve washer 'F' as shown in the illustration against the face 'D' at the outlet end of the cylinder. The sealing lip 'A' of the rubber cups in the front expanders is normally held in firm contact against the bore of the expander by a hydraulic pressure of (approx) 12-14 lbs per sq.in. created by the spring loading of the return valve. If this valve fails to maintain the hydraulic pressure air will enter past the cups of the expanders due to the low pressure of the cups against the bore.

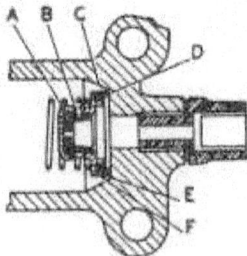

THE RETURN VALVE ASSEMBLY.

The Master Cylinder.

If the master cylinder is suspected, remove and dismantle for inspection. A situation can arise whereby the brakes appear inefficient after prolonged application, and when increased pedal pressure is applied before a final stop, the application becomes extremely hard.

This can be due to a groove worn on the top of the rubber master cup (Page G20,8,Fig.G19) caused by its travel past the small replenishing hole between the cylinder and the reservoir, causing the liquid to leak past the master cup.

The cylinder should be polished as already described and all scratch markings removed (if they are not to deep). But in cases of scoring the master cylinder and reservoir must be discarded and replaced with a new unit, and a new cup fitted.

SILVER WRAITH — SILVER DAWN — BENTLEY MK. VI.
R. TYPE BENTLEY — PHANTOM IV.

# SECTION
# H
# FRONT SUSPENSION - STEERING

# SERVICE HANDBOOK

SILVER WRAITH — SILVER DAWN — BENTLEY MK. VI.
R. TYPE BENTLEY — PHANTOM IV.

SECTION H

FRONT SUSPENSION AND STEERING

STEERING GEOMETRY - STEERING COLUMN AND BOX - SIDE AND CROSS
STEERING TUBES - CENTRE STEERING LEVER AND SWIVEL PIN HOUSING
ASSEMBLY - STUB AXLES - PIVOT PINS - YOKES - FRONT SPRINGS
- LOWER TRIANGLE LEVERS - TORQUE ARMS - FRONT SHOCK DAMPERS

# SERVICE HANDBOOK

### SILVER WRAITH — SILVER DAWN — BENTLEY MK. VI.
### R. TYPE BENTLEY — PHANTOM IV.

SECTION H

FRONT SUSPENSION AND STEERING

List of Illustrations:

SILVER WRAITH — SILVER DAWN — BENTLEY MK. VI.

R. TYPE BENTLEY — PHANTOM IV.

STEERING AND FRONT SUSPENSION

GENERAL:

The steering is of the cam and roller type. A double toothed follower roller mounted in the jaws of the rocking shaft, engages with the cam, this being a modified form of worm gear.

The independent front wheel suspension system consists of the two upper and two lower radius arms, of different lengths, set at a leading angle, with open type coil springs mounted between the forward lower radius arms and the chassis frame. Each lower radius arm consists of a front lever operating in a silentbloc rubber bearing at its inner end and a torque arm operating in a spherical rubber bearing at the rear end.

The upper radius arms also constitute the arms for the double acting hydraulic shock dampers.

Mounted between the radius arms are the vertical yokes on which the stub axles are pivoted.

A steel torsion-rod stabiliser, anti-roll bar, is provided, mounted in rubber bearings and coupled to the wheel mountings by links and rubber pads.

The front wheels and hubs being rubber insulated in relation to the rest of the chassis, it is important that the rubber bushes should be in 'a normal state of compression' when checking the geometry.

STEERING GEOMETRY:

Commencing Bentley chassis B-1-GT, Silver Wraith chassis WME-1, and Silver Dawn chassis SCA-1, the steering geometry was modified to provide greater steering accuracy. It is not possible to apply the revised design to earlier chassis.

The changes in the revised design are:-

1.  The lower yoke bearings at the outer ends of the lower triangle levers have been raised 1.500", thus reducing the angularity of the lower triangle levers by 1.100" and increasing the ground clearance. The angularity of the upper triangle levers has been similarly reduced.

    Thus, the upper triangle lever assembly of the front shock dampers is not interchangeable with the assembly fitted to earlier series chassis owing to the different angular setting of the lever in relation to the main shaft of the lever.

2.  The effective radius of the track rods and their angularity has been increased by interposing a third section between the two swinging track rods, involving the use of an extra steering lever on the front pan.

3.  The free and loaded length of the front suspension springs is reduced by .600", but the number of coils and the rating remain the same.

### SILVER WRAITH — SILVER DAWN — BENTLEY MK. VI.
### R. TYPE BENTLEY — PHANTOM IV.

4.  The off-set of the point of contact of the tyres from the projected centre line of the pivots is reduced, to reduce any disturbance of the steering should one brake to more effective than the other.

5.  The steering drop arm has been shortened to lighten the steering.

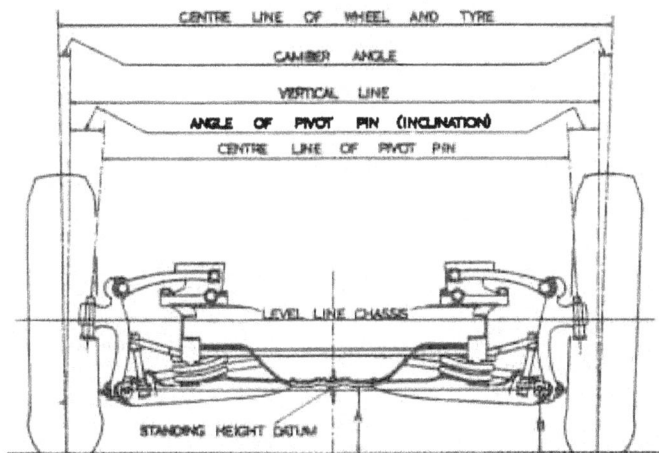

FIG.H1.  STEERING GEOMETRY.

## STANDING HEIGHT:

As shown in Fig. H1, the standing height is the difference between the dimensions "A" and "B".

To check the measurement:-

1.  The car should be unladen, i.e. less driver and passengers, but with five gallons of petrol in the tank, and standing on level ground.

2.  Check front and rear tyre pressures and correct.

3.  Measure distance from ground to the underside (centre rib) of the centre plate of the front pan - Dimension "A".

4.  Measure the distance from the ground to the centre of the lower bearing of the yoke, i.e. the centre of the bolt which passes through the bearing - Dimension "B".

5.  Subtract "B" from "A" and record the reading obtained.

**SILVER WRAITH — SILVER DAWN — BENTLEY MK. VI.**

**R. TYPE BENTLEY — PHANTOM IV.**

A negative reading is when dimension "B" is greater than "A". A positive reading is when dimension "A" is greater than dimension "B".

The limits for standing height are:-

Bentley and Silver Dawn:

| | | |
|---|---|---|
| Standard Springs | .700" to 1.350" positive | - early chassis |
| Standard Springs | .400" negative to .200" positive | - later chassis |
| Colonial Spring | 1.350" to 2.050" positive | - early chassis |
| Colonial Springs | .600" positive to 1.200" positive | - later chassis |

Silver Wraith:

| | | |
|---|---|---|
| | 1.150" to 2.150" positive | - early chassis |
| | .300" to .900" positive | - later chassis |

Phantom IV:

1.150" to 2.100" positive

Adjustment:-
No precise instructions can be laid down, each car must be individually considered in relation to the weight on the front wheels and the poundage of the springs fitted. Certain packing washers may be fitted but extra packing washers should not be added before consulting the Main Service Station. It is however, permissible to remove any packing washers to lower the car. The standing height will be reduced by twice the thickness of any washer removed.

Under-steer can be improved slightly at the expense of joggles, by raising the front of the car to the top limits of standing height.

TOE-IN OF FRONT WHEELS:
The Factory setting for toe-in on new cars is $5/32$" $- 1/16$. $^{+ 1/32}$ This allows for the settling down of the various Silentbloc bushes, and this setting should be used if these bushes are renewed. Under normal running conditions, the toe-in must be set within the limits of $1/16$" to $1/8$".

To Measure:-
Use a standard optical alignment gauge, as per the makers instructions.

Adjustment:-
The cross steering tubes are adjustable for length. One complete turn of tube and socket will alter the toe-in by .090". Disconnect outer end of cross steering tube from cross steering lever, using steel drift and steady block to release tapered shank of ball pin. Slack off the pinch bolt at inner end and unscrew to increase, divide more than one turn between both cross steering tubes.

WHEEL CAMBER:
With two passengers in the front seats of the car, the camber angle should be $1°$ to $1\frac{1}{4}°$ outwards on the earlier models and $0°$ (vertical) to $1°$ outwards on later models. Phantom IV, $0°$ to $1\frac{1}{4}°$ outwards. There is no provision for adjustment.

### SILVER WRAITH — SILVER DAWN — BENTLEY MK. VI.
### R. TYPE BENTLEY — PHANTOM IV.

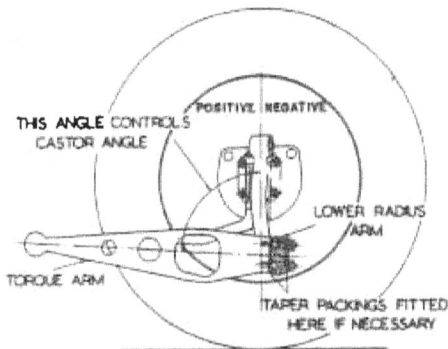

FIG. H2. CASTOR ANGLE.

CASTOR ANGLE:

The castor angle, see Fig. H2, should be $\frac{3}{4}^\circ$ positive to $1\frac{1}{2}^\circ$ negative for the earlier models, and $\frac{1}{2}^\circ$ positive to $1\frac{1}{2}^\circ$ negative for the later models. Phantom IV, vertical to $1^\circ$ negative.

Adjustment:-

The castor angle can be made positive by fitting taper wedges between the torque arm and the lower radius arm, see Fig. H2, $1^\circ$ taper wedges will probably be adequate in most cases. $1\frac{1}{2}^\circ$ wedges can be fitted if desired, but they tend to increase road shocks.

PIVOT PIN INCLINATION:

The pivot pin (king pin) inclination (angle) is approximately $3\frac{1}{4}^\circ$ on the earlier models and $4\frac{1}{2}^\circ$ on the later series. Phantom IV, is approximately $3\frac{1}{4}^\circ$. No adjustment is provided.

### THE STEERING COLUMN AND BOX

TO REMOVE STEERING COLUMN AND BOX:

(i)   Disconnect the controls at the bottom of the column, see Fig. H3. Remove the horn wire from the horn relay. Remove anti-chafing bush (6), at bottom of steering column and the pinch bolts from levers.

(ii)  Unlock and remove nut (2) from threaded taper piece (7) and remove nearside nut (8). Replace nut (2) on taper piece, tap towards the steering box to loosen tube, then slip nut over tubes and remove control tube assembly.

(iii) Remove steering wheel using Extractor No. 3243/T1006. Mark hub of steering wheel and steering cam for re-assembly.

- H.4 -

### SILVER WRAITH — SILVER DAWN — BENTLEY MK. VI.
### R. TYPE BENTLEY — PHANTOM IV.

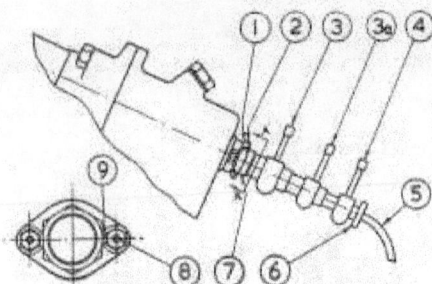

SECTION 'AA'

FIG. H3.   STEERING COLUMN (BOX) - Lower End.

1. Locking Washer.
2. Nut.
3. Throttle Control Lever.
3a. Riding Control Lever.
4. Mixture Control Lever, on
   Bentley cars (prior to Series "R").

5. Horn Wire.
6. Anti-chafing Bush.
7. Taper Piece (Threaded).
8. Nut.
9. Locking Tab.

(iv)   Remove rubber gas seal (around steering column), detach seal from
       dashboard.

(v)    Remove front R.H. undershield, and, jack up and remove the side
       steering tube (17, Fig. H4) from the pendulum lever (21) (see page H12).

       If steering is turned on full R.H. lock, ball pin will come clear
       of the pendulum lever.  If not, it will free when the steering box
       and bracket are disconnected from the frame.

15. Pressure Spring.
16. Sealing Disc.
17. Side Steering Tube.
18. Mud Excluder.
19. Ball Pin (Tapered).
20. Locating Washer.
21. Pendulum Lever.

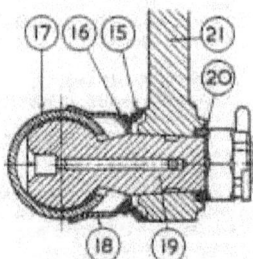

FIG. H4.   SECTION THROUGH
PENDULUM LEVER BALL PIN.

(vi)   Disconnect steering box complete with its bracket (42, Fig. H5).

(vii)  Disconnect oil feed pipe to the ball pin, from 4 way connection.

### SILVER WRAITH — SILVER DAWN — BENTLEY MK. VI.
### R. TYPE BENTLEY — PHANTOM IV.

(viii) While steering column is held, remove bracket under the instrument panel. Remove the column and box, by passing it forwards.

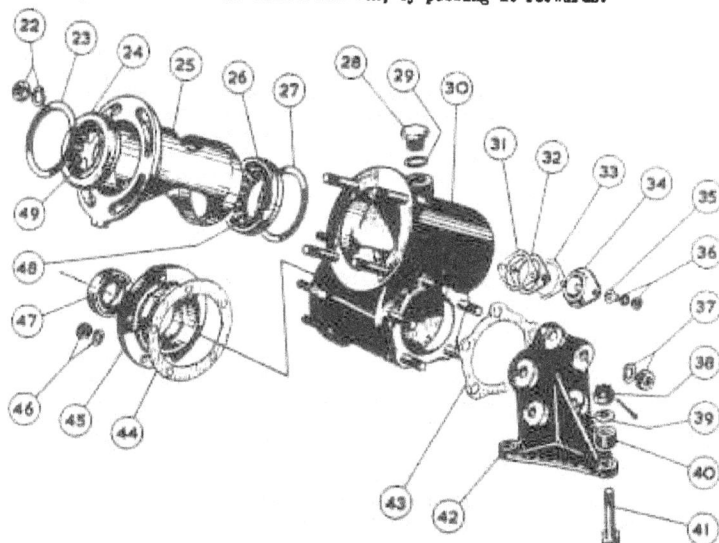

FIG. H5. STEERING BOX, "EXPLODED VIEW".

| | |
|---|---|
| 22. Nut and Spring Washer. | 36. Nut and Spring Washer. |
| 23. Adjusting Washer (Range of). | 37. Nut and Spring Washer. |
| 24. Roller Bearing Cup. | 38. Nut. |
| 25. Adjusting Sleeve. | 39. Plain Washer. |
| 26. Roller Bearing Cup. | 40. Distance Piece. |
| 27. Adjusting Washer (Range of). | 41. Bolt. |
| 28. Oil Filler Plug. | 42. Bracket. |
| 29. Washer. | 43. Joint Washer. |
| 30. Steering Box. | 44. Joint Washer. |
| 31. Joint Washer. | 45. Cover. |
| 32. Adjusting Washer (Range of). | 46. Nut and Spring Washer. |
| 33. Joint Washer. | 47. Oil Seal. |
| 34. Guide. | 48. Roller and Cage Assembly. |
| 35. Locking Tab. | 49. Roller and Cage Assembly. |

To Dismantle Steering Column and Box:

Drain oil from steering box. Hold column in a vice at lower end near the box, using wood vice clamps. The box itself MUST NOT be held in vice, mark pendulum lever and rocker shaft for re-assembly.

Remove pendulum lever and coiled pipe from rocking shaft using Extractor No. 3243/T1001.

- H.6 -

Leakage or wear of the oil seal (47, Fig. H5) will be indicated by oil inside the cover and on the pendulum lever.

Removing Rocker Shaft from Steering Box:

(i)   Mark cover (45, Fig. H5) and steering box for re-assembly. Remove cover and oil seal (47) and roller bearing (59, Fig. H6). Rocking Shaft can be removed with roller and cage assembly (55).

(ii)  Remove nuts retaining steering column (67, Fig. H7) to box, and remove steering column.

(iii) Remove the nut securing stationary tube guide (34, Fig. H5) to the box, remove the guide with an aluminium drift or screwdriver. A joint washer is fitted between guide and box. An adjusting washer/s (32) and an extra joint washer is fitted between the guide and the box (as shown in Fig. H5) to control the end play of the stationary tube (64, Fig. H7).

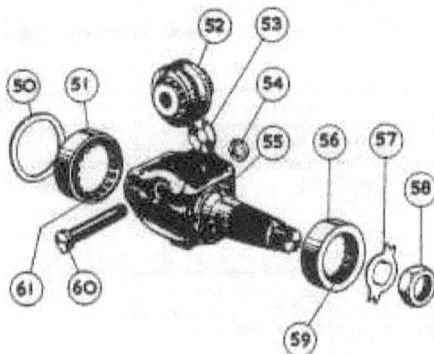

FIG. H6.  ROCKING SHAFT ASSEMBLY, "EXPLODED" VIEW.

50.  Adjusting Washer (Range of).
51.  Roller Bearing Cup.
52.  Cam Roller Assembly.
53.  Adjusting Washer (Range of).
54.  Nut.
55.  Rocking Shaft.
56.  Roller Bearing Cup.
57.  Lock Washer.
58.  Nut.
59.  Roller and Cage (Assembly).
60.  Bolt.
61.  Roller and Cage (Assembly).

SILVER WRAITH — SILVER DAWN — BENTLEY MK. VI.

R. TYPE BENTLEY — PHANTOM IV.

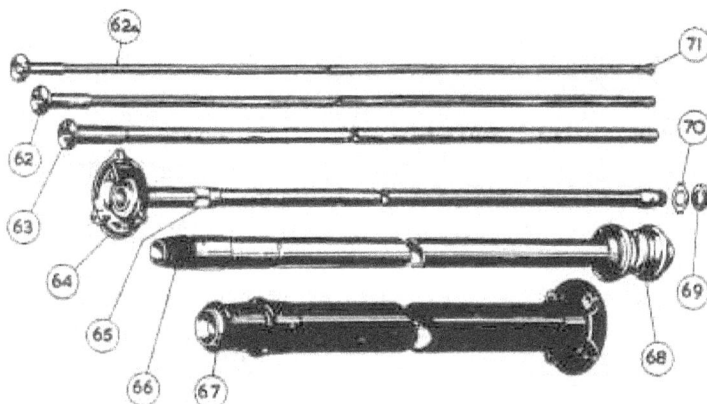

FIG. H7. STEERING COLUMN, "EXPLODED" VIEW.

| | |
|---|---|
| 62a. Mixture Control Tube (Assembly) (Bentley cars). | 66. Steering Cam Tube Assembly. |
| 62. Riding Control Tube Assembly. | 67. Steering Column Assembly. |
| 63. Throttle Control Tube Assembly. | 68. Steering Cam. |
| 64. Stationary Tube Assembly. | 69. Nut. |
| 65. Felt Packing Strip. | 70. Lock Washer. |
| | 71. Anti-chafing Bush. |

(iv)  Using an aluminium drift on the bottom end of the steering cam tube assembly, at the same time, pull the cam tube by hand to withdraw it from the steering box. The cam tube assembly, on removal from the box, may withdraw the adjusting sleeves (25, Fig. H5), if so, remove it by tapping its upper face until the upper bearing cup (24) of the taper roller bearing is clear of the sleeve.

(v)  Thoroughly clean all parts.

To Remove Cam Roller Assembly from the Rocking Shaft:

It should only be necessary to remove the cam roller assembly (52, Fig. H6) from the rocking shaft in the event of wear, end float of the assembly in the rocker shaft, or a flat on the cam track due to shock.

(i)  Remove bolt (60), cam roller assembly (52) and adjusting washer/s (53) from rocking shaft.

(ii)  Thoroughly clean all parts.

To Fit a New Cam Roller Assembly to the Rocking Shaft:

(i)  Determine the thickness of the adjusting washer (53, Fig. H6) to ensure contact of the inner end faces of the two inner races of the cam roller assembly, to give the necessary pre-loading on the roller.

# SERVICE HANDBOOK

### SILVER WRAITH — SILVER DAWN — BENTLEY MK. VI.
### R. TYPE BENTLEY — PHANTOM IV.

The inner race of the cam roller is in two halves, (52, Fig. H6) held together by a retaining ring. When the inner faces of the two halves of the inner race are brought into contact, this will give a pre-load on the roller of 2 to 8 ounces at a radius of 3" the roller will be a little stiff to turn by hand. This pre-load is determined by the makers and cannot be altered.

(ii)   Before fitting cam roller assembly to rocking shaft, place on bolt (60, Fig. H6), a temporary distance piece 7/16" thick by about 11/16" outside diameter by 7/16" inside diameter, or failing this, suitable washers, to compensate for the thickness of the walls of the rocking shaft. Pass the bolt through the inner races and then tighten up to bring into contact the inner faces of the two halves of the inner race.

With nut and bolt, tightened, measure the width across the outer faces of the inner races, which should be 1.257" plus or minus 0.003", then measure the width across the inner machined faces of the rocking shaft gap which will be 1.262" plus 0.005". The difference between the measurements will determine the thickness of the adjusting washer required.

(iii)   Fit the roller assembly into the rocking shaft ensuring that the adjusting washer makes the roller assembly a good push fit in the rocker shaft gap. Insert the bolt. Check roller for freedom. It should be a little tight. If free, the adjusting washer/s are not thick enough to ensure contact of the inner races.

## To Fit Rocking Shaft to Steering Box, Temporarily in Order to Check Pre-Load of Rocking Shaft:

(i)   Place one of the two roller and cage assemblies (59 or 61, Fig. H6) on to the bearing cup (51), with the bearing cup in the steering box and drop the rocking shaft into position in the box, Place the other roller and cage assembly on the rocking shaft. Place the cover with bearing cup (56) and oil seal over the splined end of the rocking shaft, with markings in line. Fit spring washers to the three short studs and temporary additional washers to the two long studs, and tighten up.

(ii)   Slip the pendulum lever on to the splines of the rocking shaft with the marking on lever and shaft in line. Hold steering box (bracket 42), in a vice, with rocking shaft and pendulum lever horizontal, check the pre-load at the end of the pendulum lever which should be 3 to 12 ounces.

To carry out this check:- Six 2 ounce weights are required, with a small hole in the centre and a piece of hooked wire for attachment. Find the weight required to move the pendulum lever, by adding weights to the end of the lever. If less than 3 ounces, fit a thicker adjusting washer as follows:-

(iii)   Remove bracket (42), joint washer (43) Fig. H5, and adjusting washer/s (50, Fig. H6).

Measure the thickness of the adjusting washer/s removed, and select a washer/s 0.001" thicker from the range available. Such a change, i.e. 0.001" will alter the pre-load by approximately 8 ounces.

- H.9 -

Place adjusting washer/s against the outer face of the bearing cup
(51, Fig. H6). Fit a new joint washer (43) and refit the bracket (42)
to the steering box with markings in line and re-check the pre-load.

To Centralise the Steering Cam with Rocking Shaft:

(i)    Should the cam tube assembly have withdrawn with it the adjusting
sleeve (25, Fig. H5), proceed as follows:-

    (a)  Grease the two outer diameters of the adjusting sleeve (25) and
enter it into position in the steering box, so that the two faces
make contact. Grease the original lower adjusting washer (27)
and with the steering box in a vertical position, place in position
in the bottom of the box. If two adjusting washers were fitted,
the thinner of the two washers must be fitted adjacent to the lower
face of the lower bearing cup (26), if the thinner washer was fitted
behind the thicker washer, there would be a risk of it being trapped
behind the gap which exists between the inner end of the adjusting
sleeve and the lower inner end of the steering box.

    (b)  Fit the lower bearing cup (26) in position against the adjusting
washer/s.

(ii)    Place one of the two roller and cage assemblies (48 or 49) on the lower
bearing cup, place cam tube assembly on the roller bearing. Place the
other roller and cage assembly on to the cam, and push or tap the other
bearing cup (24) squarely into position in the adjusting sleeve and on
to the roller bearing. Select an adjusting washer/s (23), too thick to
enable the flange of the steering column to be pulled right down on to
the adjusting sleeve. Place the adjusting washer/s against the outer
face of the upper bearing cup (24) and then place steering column in
position, nip up the steering column retaining nuts sufficiently to make
the cam tube moderately stiff to turn by hand before the steering wheel
is slipped on.

There should now be a gap between the face of the flange of the steering
column and the adjusting sleeve. It is possible for the adjusting sleeve
to "lift", creating the gap between the flange and sleeve - avoid this,
otherwise the lower adjusting washer/s (27) could slip out of position
and become trapped. If there is no gap then a thicker adjusting washer/s
(27) must be fitted until a gap is obtained.

(iii)  The eccentric adjusting sleeve is for adjusting the mesh between the cam
and roller. Looking from the steering column end, rotate the sleeve
clockwise to "free" the cam from the roller, i.e., there will be slackness
between the cam and roller. The sleeve has elongated stud holes which
limit its rotation in the box. Rotate the sleeve by the lug provided -
clockwise, "free" the cam from the roller and nip up the steering column
nuts again.

Slacken steering column nuts just sufficiently to allow the adjusting
sleeve to be rotated by the lug.

(iv)    Temporarily refit rocking shaft to steering box and secure in position
by refitting the cover (45, Fig. H5).

SILVER WRAITH — SILVER DAWN — BENTLEY MK. VI.

R. TYPE BENTLEY — PHANTOM IV.

Mount steering box in a vice and slip on the steering wheel. From one full lock position, rotate slowly towards the other lock, hold the end of the pendulum lever with the other hand, shaking it to and fro until the position of minimum slack between the cam and roller has been found. With the cam in this position and without moving it, again slacken the four steering column nuts slightly, i.e. only just sufficiently to allow the adjusting sleeve to be rotated. Rotate sleeve anti-clockwise until all slack between the cam and roller is just eliminated, then retighten.

(v)    Set cam in the straight.ahead position, by rotating the steering wheel about one-and-seven-eighth's of a turn from either lock, bringing the keyway in the cam tube to its lowest position, i.e. nearest to the rocking shaft.

The tightest meshed position extends over possibly half a turn of the steering wheel. By rotating the steering wheel, determine how much it has to be turned to move it from the straight ahead position to the centre of the tightest meshed position. If the tightest meshed position is to the right of the straight ahead position, a thinner adjusting washer/s (27, Fig. H5) should be fitted to the lower roller bearing; if to the left, then a thicker washer is required. If two washers are fitted, the thinner must be fitted against the lower face of the bearing cup. The washer/s should be greased to keep them in position against the cup during erection. To estimate the thickness of the washer required, a change of 0.007" moves the tight place along half a turn of the steering wheel. When the centre of the tightest meshed place is at the straight ahead position, i.e. cam tube assembly keyway at its lowest point, and rocking shaft in central position, adjust the pre-load on the cam bearings.

## To Adjust Pre-Load on Cam Roller Bearings:

(i)    The pre-load should be from 12 to 20 ounces, measured at the rim of the steering wheel, and this should be done with the column in a horizontal position. The pre-load is obtained by selecting a washer/s to fit behind the lowest face of the steering column flange and the outer face of the upper bearing cup (24).

(a)    Measure the gap at two opposite points between the flange faces of the adjusting sleeve and the steering column flange and note the average gap. Loosen the nuts retaining the column to the box and rotate the adjusting sleeve (25), clockwise producing slack between the cam and roller.

(b)    Remove column and adjusting washer/s (23). To determine the thickness of the adjusting washer/s required, measure the thickness of the washer/s removed, and subtract from this figure, the average gap.

(ii)   Fit the new washer/s as determined, replace the steering column on the box. Slip on the steering wheel and attach a piece of string to the steering wheel rim, to this, tie on 2 ounce weights as required to determine the pre-load. If the weight required to move the steering wheel exceeds 20 ounces, decrease the thickness of the upper adjusting washer/s by 0.001", but if below 12 ounces, increase the thickness of the washer/s by 0.001".

## SILVER WRAITH — SILVER DAWN — BENTLEY MK. VI.
## R. TYPE BENTLEY — PHANTOM IV.

TO REFIT STEERING COLUMN AND BOX:

(i)     If an adjusting washer/s (32) was fitted between the guide (34) and the
        box (as shown in Fig. H5), fit a new joint washer (31 and 33) either
        side of the adjusting washer/s. Secure the guide to the box by placing
        one of the two spring washers and nuts on the off-side stud and tighten
        up. Do not fit the other nut (8, Fig. H3).

(ii)    With the coiled lubricating pipe cover on the pendulum lever, fit the
        lever to the rocking shaft in its original position with markings in
        line, i.e. pendulum lever $\frac{1}{4}°$ backwards. Ensure alignment of the oil
        holes in the ball of the pin with the axis of the side steering tube.

To Check and Adjust Controls:

        The throttle control lever (3, Fig. H3), the ride control lever (3a)
and the mixture control lever (4) (on Bentleys prior to series "R"), may not have
been refitted in the same angular position relative to the tube, the following
adjustments should be made if necessary:-

        Hand Controls::

        The hand throttle lever on steering column should move from fully
closed position, approximately $\frac{3}{4}$" before starting to open the throttle.
If free travel is more or less than $\frac{3}{4}$", set lever in the fully closed
position, slacken pinch-bolt and ease the lever as required. The other
controls can be adjusted in a similar manner.

### THE SIDE AND CROSS STEERING TUBES

SIDE STEERING TUBE:

        The ball joints are lubricated from the centralised chassis lubrication
system. No adjustment is provided for the poundage on the ball joint at the front
end of the tube, as a coil spring exerts a constant pressure on the joint (see
Fig. H9), it can be adjusted for length. The length is fixed by the makers during
the building of the chassis; it is screwed into a socket and attached to the
centre steering lever. If replaced, the new tube must be adjusted for length.

To Remove a Side Steering Tube:

Front End:

(i)     Remove nut (12, Fig. H8) from the ball pin at the front end of tube.
        Remove ball pin (8), pressure spring (13) and sealing disc (11).

Rear End:

(ii)    Remove the side steering tube (ball pin) from the pendulum lever.

(iii)   If there is insufficient clearance to allow the ball pin to come clear
        of the pendulum lever, disconnect coil pipe and remove the pendulum
        lever from the rocking shaft using Extractor No. 3243/T1001.

        Mark the pendulum lever and rocking shaft for re-assembly.

SILVER WRAITH — SILVER DAWN — BENTLEY MK. VI.

R. TYPE BENTLEY — PHANTOM IV.

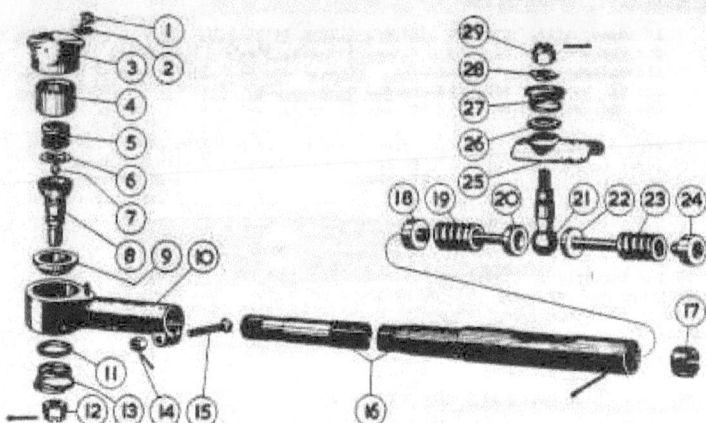

FIG. H8.   SIDE STEERING TUBE, "EXPLODED" VIEW.

| | | | |
|---|---|---|---|
| 1. | Nut and Spring Washer. | 16. | Side Steering Tube. |
| 2. | Locking Plate. | 17. | End Nut (Internal). |
| 3. | Cap Nut. | 18. | Guide. |
| 4. | Sleeve. | 19. | Spring. |
| 5. | Spring. | 20. | Ball Pad. |
| 6. | Spring Pad. | 21. | Ball Pin (Tapered). |
| 7. | Ball, hardened 0.3125" dia. | 22. | Ball Pad. |
| 8. | Ball End Pin. | 23. | Spring. |
| 9. | Ball Pad. | 24. | Guide. |
| 10. | End Socket. | 25. | Mud Excluder. |
| 11. | Sealing Disc. | 26. | Sealing Disc. |
| 12. | Nut. | 27. | Pressure Spring. |
| 13. | Pressure Spring. | 28. | Locating Washer. |
| 14. | Nut. | 29. | Nut. |
| 15. | Pinch Bolt. | | |

To Remove Ball End Pin from Front End of Side Steering Tube:

(i)     Remove nut, spring washer (1), Fig. H8, and locking plate (2);
Cap nut (3) and sleeve (4);  spring (5), spring pad (6), ball (7)
and ball pin (8).  (The ball pad (9) can be removed if necessary).

(ii)    When re-assembling, pack cap with Duckham's HBB. Grease.

To Remove Ball End Pin from Rear End of Side Steering Tube:

(i)     Remove split pin and internal end nut (17), guide (24), spring (23)
ball pad (22), ball pin (21);  then ball pad, spring and guide.

(ii)    Clean all parts thoroughly.

To Refit Ball End Pin to Rear End of Side Steering Tube:

Smear parts with oil and re-assemble in reverse order of dismantling.
Screw in the end nut (17) until the ball pads, spring and guide
assemblies are choc-a-bloc. Slacken the nut back 0.180" to restore
the original standard working clearance of 0.090" at points 'A' and
'B', Fig. H9.

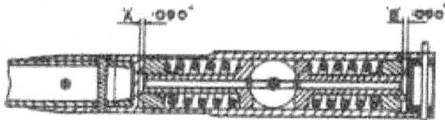

FIG. H9. SECTION - BALL JOINT AT REAR END OF
SIDE STEERING TUBE.

To Refit the Side Steering Tube:

Rear End:

Ensure that the mud excluder (25, Fig. H8) is fitted correctly.
Enter the pin (21) into the pendulum lever. Place the locating washer (28)
on to the ball pin to ensure alignment of the oil holes in the ball of the
pin with the axis of the side steering tube. If the pendulum lever has been
removed from the rocking shaft, refit the ball pin of the side steering tube
to the pendulum lever on the bench. With the coiled lubrication pipe and cover
in position on the pendulum lever, fit lever to rocking shaft in original
marked position.

Front End:

Place the spherical face of the sealing disc (11) against the ball
pad (9, Fig. H8), place the small diameter end of the pressure spring (13)
on to the spigotted sealing disc. Enter the ball pin into the centre steering
lever.

To Adjust a New Side Steering Tube for Correct Length:

(a)  Rotate steering wheel about one and seven eighths of a turn from
     either lock so as to place the cam tube in the straight-ahead
     position (i.e. the cam roller of the steering, central with the
     cam). The spoke nearest to the oil hole in the hub of the steering
     wheel should now be at the top. Place the spoke at the top as
     necessary, so as to get the cam tube in the exact straight-ahead
     position.

(b)  With the road wheels in the straight-ahead position and, without
     moving either steering wheel or front wheels, remove pinch bolt (14)
     from side steering tube and adjust for length by screwing end socket
     (12) in or out as required. Refit pinch bolt.

SILVER WRAITH — SILVER DAWN — BENTLEY MK. VI.

R. TYPE BENTLEY — PHANTOM IV.

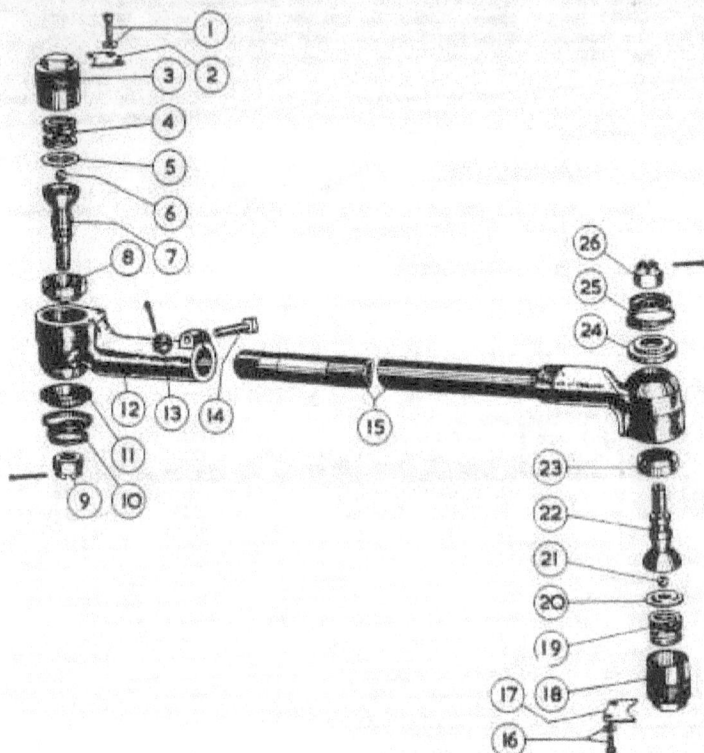

FIG. H10. CROSS STEERING TUBE, "EXPLODED" VIEW.

| | | | |
|---|---|---|---|
| 1 & 16. | Setscrew and Spring Washer. | 9 & 26. | Nut. |
| 2 & 17. | Locking Plate. | 10 & 25. | Pressure Spring. |
| 3 & 18. | Cap Nut. | 11 & 24. | Sealing Disc. |
| 4 & 19. | Spring. | 12. | Socket (Inner). |
| 5 & 20. | Spring Pad. | 13. | Nut. |
| 6 & 21. | Ball, Hardened, 0.3125" dia. | 14. | Pinch Bolt. |
| 7 & 22. | Ball End Pin. | 15. | Cross Steering Tube and Outer Socket. |
| 8 & 23. | Ball Pad. | | |

SILVER WRAITH — SILVER DAWN — BENTLEY MK. VI.

R. TYPE BENTLEY — PHANTOM IV.

CROSS STEERING TUBES:

The four ball joints are identical and lubricated from the centralised chassis lubrication system. A coil spring exerts a constant pressure on the joints. (See Fig. H9). In the event of wheel wobble or of road re-action transmitted to the steering wheel, increase the loading of the two outer ball joints. To increase the loading, fit a packing washer or shim between the coil spring (19, Fig. H8) and the cap nut (18). Ensure that the thickness of the packing washer is such that when the cap nut is fully screwed up, the spring is not choc-a-bloc. The cross steering tubes are adjustable for front wheel alignment (toe-in) purposes as described in "Steering Geometry".

To Remove a Cross Steering Tube:

Remove split pin and nut (26, Fig. H10) from ball pin (22) and remove ball pin from the lever. Collect pressure spring (25) and sealing disc (24).

To Remove a Ball End Pin from a Joint:

(i) Remove setscrews, spring washers (1, Fig. H10) and locking plate (2).

(ii) Remove cap nut (3), spring (4), spring pad (5), ball (6) and ball pin (7). The ball pad (8) can be removed if necessary.

(iii) Clean all parts thoroughly. (When re-assembling, pack with Duckham's HBB. grease).

## THE CENTRE STEERING LEVER AND SWIVEL PIN HOUSING ASSEMBLY

LUBRICATION:

Oil under pressure from the centralised chassis lubrication system, is delivered to the drilled top cover, (6, Fig. H11 and 1, Fig. H12). It then passes down the passage in and through two small diameter holes in the stem of the spring loaded restrictor, (35, Fig. H11 and 34, Fig. H12), to lubricate the upper and lower bushes. The clearance being arranged to meter the correct amount.

The swivel pins (25, Fig. H11 and 24, Fig. H12), are also drilled with two small holes which register on annular grooves with the two small oil holes in the restrictor, and an oil passage in the centre steering lever. The oil is then carried to the inner ball joints of the cross steering tubes and the ball joint at the front end of the side steering tube.

Oil leakage from the centre steering lever joints is probably due to expansion of air in the joints after operation of the chassis oil pump. The cure is to dismantle the ball joints and to pack them with Duckham's HBB. grease. The chassis oil pump develops sufficient pressure to clear the oil feed pipe.

Nomenclature for Fig. H11:

| | | |
|---|---|---|
| 1. Bolt (Special). | 19. Union Nut. | 40. Felt.Washer. |
| 2 & 3. Bushes. | 22 & 23. Ball Pins. | 41. Adjusting Shim (Range). |
| 4. Junction (Straight). | 25. Swivel Pin. | X.24. Nut,Spring & Plain |
| 5. Top Cover. | 28. Spring Seat. | Washer. |
| 10. Housing (Lower). | 30. Compression Sleeve. | X.50. Screw. |
| 11. Housing (Upper). | 32. Spring Valve Restrictor. | X.57. Bolt. |
| 14. Centre Steering Lever. | 33. Spring. | X.84. Plain Washer. |
| 17. Cross Tube Centre Link. | 35. Restrictor Valve. | X.87. Spring Washer. |

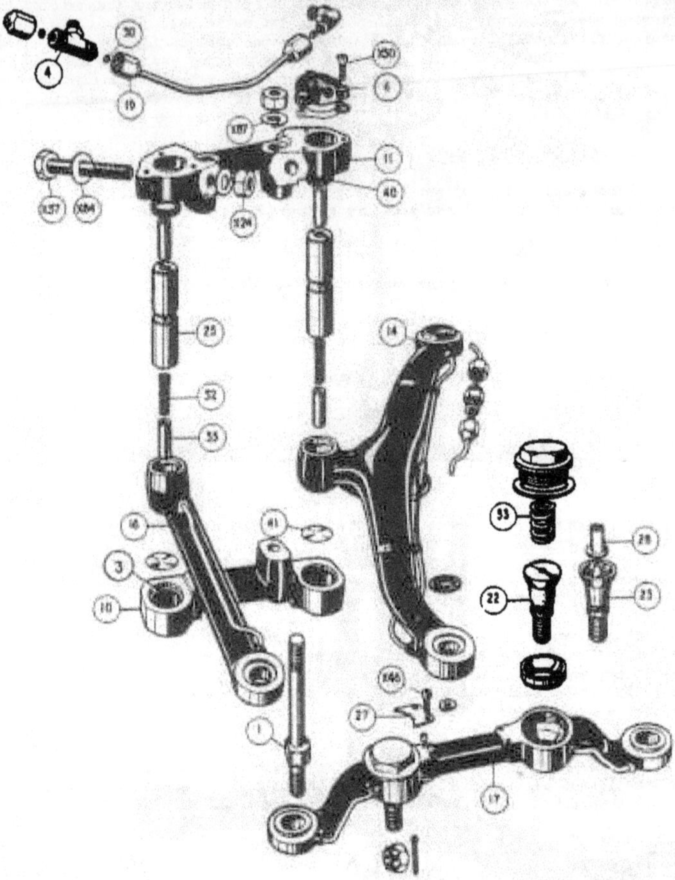

SILVER WRAITH — SILVER DAWN — BENTLEY MK. VI.

R. TYPE BENTLEY — PHANTOM IV.

FIG. H11. "EXPLODED" VIEW OF SWIVEL PIN HOUSING AND CENTRE STEERING LEVER NEW GEOMETRY FROM BENTLEY CHASSIS NO. B-1-GT, SILVER WRAITH WME-1, AND SILVER DAWN SCA-1.

SILVER WRAITH — SILVER DAWN — BENTLEY MK. VI.

R. TYPE BENTLEY — PHANTOM IV.

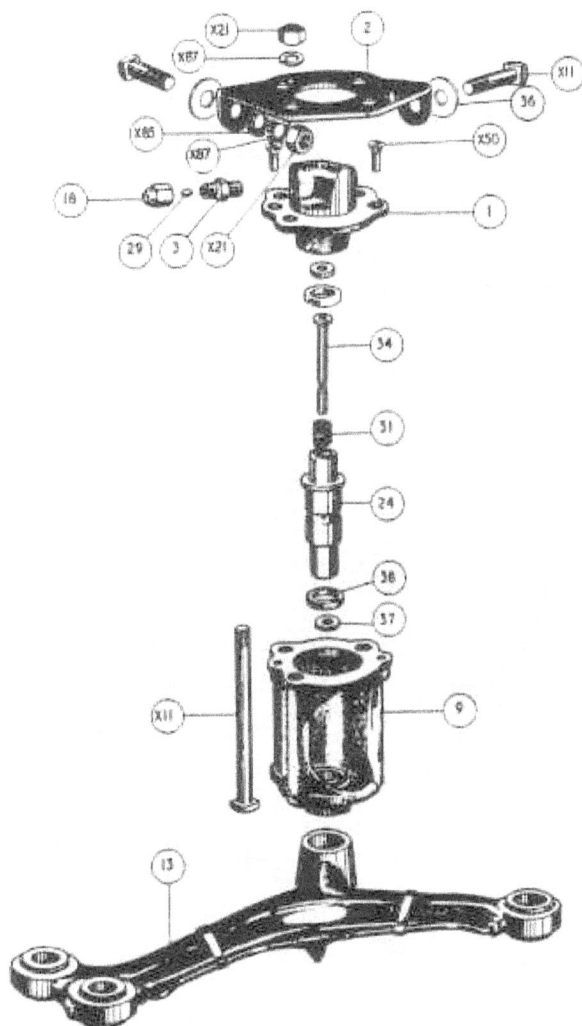

FIG. H12. STEERING LEVER FITTED PRIOR TO BENTLEY CHASSIS NO. B-1-GT, SILVER WRAITH WGE-1, and SILVER DAWN SCA-1.

SILVER WRAITH — SILVER DAWN — BENTLEY MK. VI.

R. TYPE BENTLEY — PHANTOM IV.

Nomenclature for Fig. H12:

| | | | |
|---|---|---|---|
| 1. | Upper Cover and Bush. | 34. | Lubrication Valve. |
| 2. | Bracket Housing to Frame. | 36. | Bracket Adjusting Washer. |
| 3. | Oil Pipe Junction. | 37. | Felt Washer. |
| 9. | Housing and Bush. | 38. | Swivel Pin Adjusting Washer. |
| 13. | Steering Lever. | X.11. | Bracket Bolts. |
| 18. | Oil Pipe Junction Nut. | X.21. | Bracket Nuts. |
| 24. | Swivel Pin. | X.50. | Screw, Housing Cover. |
| 29. | Compression Sleeve. | X.85. | Washer, Frame Bolt. |
| 31. | Lubrication Valve Spring. | X.87. | S. Washer, Frame Bolt. |

## To Remove a Swivel Pin Housing and Centre Steering Lever Assembly:

(i)    Remove the ball joint of side steering tube from centre steering lever, and the two inner ball joints of cross steering tubes from the cross tube centre link.

(ii)    Disconnect the oil feed pipe from top cover, remove nuts and spring washers from lower end of retention bolts (1) Fig. H11, remove the two horizontal bolts (X.57) in upper housing and withdraw assembly.

(iii)    Remove inner ball pins of cross tube centre link (17) Fig. H11, and the three countersunk screws (X.50) from top cover, lift off cover, then remove nuts from upper end of the two retaining bolts (1) and lift off upper bearing housing.

(iv)    Remove centre steering lever (14) and idler lever (16) from the lower bearing housing (10), then remove the oil restrictor valves and springs from the two levers (14 and 16).

(v)    Remove the shims (41) from inside of lower bearing housing bushes (10) and thoroughly clean all parts.

## To Re-Assemble the Swivel Pin Housing and Centre Steering Lever:

(i)    Ensure that the swivel pins are a tight fit in the steering levers. The diameter of the bore of the lever should be 0.750" + 0.0005" and the corresponding diameter of the swivel pin should be 0.751" + 0.0005". Temporarily re-assemble the unit minus the felt washers, restrictor and spring, check the amount of swivel pin end float. Fit adjusting shims as necessary to allow from 0.001" to 0.002" end float.

(ii)    Renew felt washers if necessary. Lubricate parts with engine oil. Place lower adjusting shim in position in casing. Fit felt washers in casing. Enter swivel pin into casing. Place spring and restrictor valve in swivel pin. Lightly smear (with a soft type of grease), upper adjusting shim and place it in top cover. Fit felt washer in top cover. Enter top cover on two swivel pins. Secure top cover to casing by means of countersunk screws.

SILVER WRAITH — SILVER DAWN — BENTLEY MK. VI.

R. TYPE BENTLEY — PHANTOM IV.

### FRONT SUSPENSION

#### STUB AXLES, PIVOT PINS, YOKES, LOWER TRIANGLE LEVERS AND TORQUE ARMS

**LUBRICATION:**

Oil under pressure from the centralised chassis lubrication system fills a "reservoir" in the pin. A dowel screw positions the pivot pin in relation to the oil passage in the yoke. Oil from the "reservoir" is supplied to the upper roller or needle bearing through the clearance between a loose restricting pin and the internal bore in the upper end of the pivot pin.

**To Remove Stub Axle from Yoke:**

(i)    Disconnect front brake hydraulic pipe from bracket end, <u>NOT from expander end</u>. Remove front hub complete with brake drum from stub axle. Remove the brake carrier plate complete with brake shoes.

(ii)    Remove the ball joint (ball pin) of the cross steering tube from cross steering lever attached to stub axle.

(iii)    Remove top cover (8 and 9, Fig. H13 and H14) and restricting pin. Remove the nut (X.24) from pivot pin, using box spanner, 1660/T1005. Remove the oil union from flange (10 and 11). Remove lower cover and thrust washer (25).

(iv)    Drive pivot pin downwards with an aluminium drift. The bottom end of pivot pin is threaded internally (R.H.) 11/16" 16 T.P.I. for extraction purposes. Remove 34 needle rollers from outer race (4) of the lower bearing, and remove stub axle from yoke. Remove oil trough (17) from stub axle (old type), felt washer and packing ring sub-assembly (21 and 16), from the yoke. Thoroughly clean all parts.

**To Remove Upper and Lower Outer Races from Stub Axle:**

Using a length of tube with faces as a drift, tap out lower outer race from stub axle, then remove the upper outer race in a similar manner.

**To Fit New Lower Outer Race to Stub Axle:**

With the bores in the stub axle, clean and free from scores or burrs, remove the tubular nut and sleeve from tool 1660/T1007, place the head (attached to the draw bar) of the tool on to the three studs at the top of the stub axle, and push it fully home. Lubricate the new outer race and place it in position on the draw bar followed by the sleeve of the tool and with the tubular nut and a tommy bar, screw up the tubular nut until the outer race is pressed fully home against the shoulder in the stub axle.

**To Fit New Upper Outer Race to Stub Axle:**

Temporarily fit the pivot pin to the stub axle on the bench.

(i)    Grease the roller path of the outer race (4) of the lower bearing sufficiently to hold the rollers in position. Place the 34 new needle rollers in the bearing and fit the pivot pin in position. Place the thrust washer (25) in recess in the bottom cover (10) and temporarily fit the bottom cover to keep the pivot pin in position.

SILVER WRAITH — SILVER DAWN — BENTLEY MK. VI.

R. TYPE BENTLEY — PHANTOM IV.

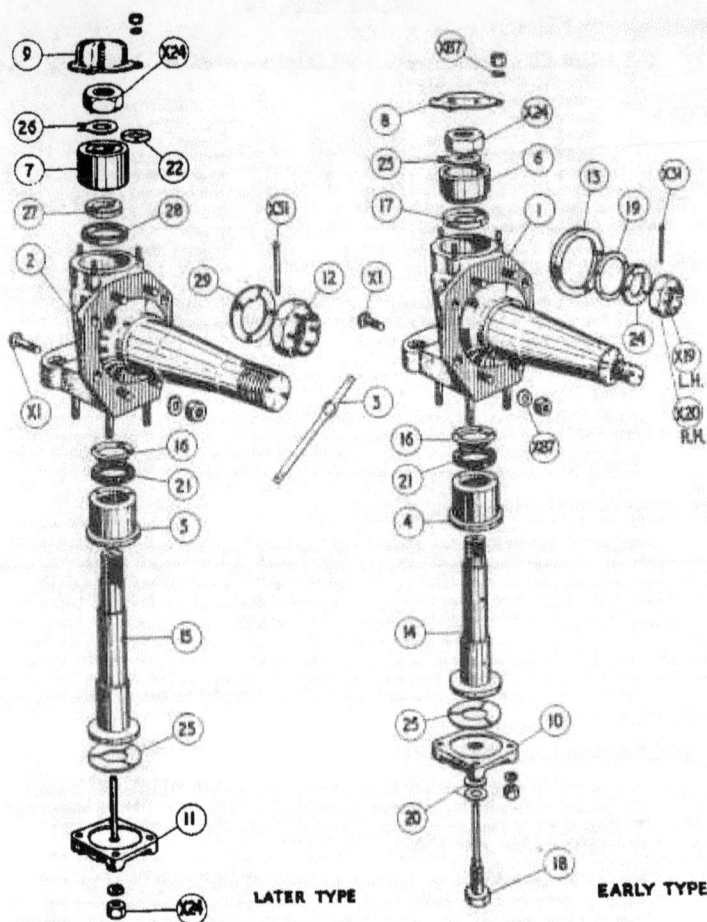

LATER TYPE

EARLY TYPE

FIG. H13. PIVOT PIN AND BEARINGS

Nomenclature for Fig. H13:

| | |
|---|---|
| 1. Stub Axle (early type). | 16. Felt Washer Retainer (Packing Ring). |
| 2. Stub Axle (later type). | |
| 3. Sealing Band. | 17. Oil Trough. |
| 4. Lower Pivot Bearing. | 18. Oil Union. |
| 5. Lower Pivot Bearing (later type). | 19. Washer. |
| 6. Upper Pivot Bearing. | 20. Washer Aluminium. |
| 7. Upper Pivot Bearing (later type). | 21. Washer Felt. |
| 8. Cover. | 22. Washer Plain (later type). |
| 9. Cover (later type). | 23. Lock Washer. |
| 10. Flange (Bottom Cover). | 24. Washer Plain. |
| 11. Flange (later, Bottom Cover). | 25. Washer Thrust. |
| 12. Castellated Nut (later type). | 26. Washer Tab. (later type). |
| 13. Distance Piece. | 27. Washer Distance (later type). |
| 14. Pivot Pin (With internal restrictor). | 28. Washer Sealing Felt. |
| | 29. Washer Special. |
| 15. Pivot Pin (later type, with internal restrictor). | X.19. Castellated Nut (later type). |
| | X.24 Pivot Nut. |

(ii)    Screw the adaptor of tool No. STD.529 on to the pivot pin. Lubricate the new outer race of the upper bearing, and with the race track at the bottom, place on adaptor. Using the special punch (STD.529), press or tap in the race until it contacts the shoulder in the stub axle.

## UPPER ROLLER BEARING ASSEMBLY (PIVOT PIN):

Whenever the stub axles have to be removed from an earlier model for reconditioning, inspect the inner roller race fitted to the upper end of the pivot pin for correct height in relation to its outer race. A condition can arise whereby the lower end of the rollers (A, Fig. H14) protrudes below the lower edge of the outer race, i.e. at point 'B' due to an accumulation of adverse limits of machined parts. Fig. H15, shows the correct position of the rollers, i.e. the lower end face of the rollers are well above the lower edge of the outer race. This applies to chassis numbers bearing the suffix letters AK to FU, in the case of Bentley cars; WTA to WLE in the case of the Silver Wraith cars; and SBA for the Silver Dawn.

## To Check, Proceed as follows:-

(i)    Having fully tightened nut at the top of pivot pin, remove nut and lockwasher beneath it, leaving pivot pin in position. It is essential to ensure that the upper outer race is fully home in its downward direction in the stub axle.

(ii)    Place a straight-edge across the top face of the outer race of the upper roller bearing assembly of the pivot pin (not across the top face of the stub axle). With a small depth gauge, measure the distance downwards from the top face of the outer race to the top face of one of the rollers (not the brass roller cage), i.e. dimension 'D' Fig. H15 and note.

(a)    If dimension 'D' is within the limits of .430" - .480" (10.9 - 12.2 m/m) this is correct.

(b) If dimension 'D' is between the limits of .480" - .515" (12.2 - 13 m/m), then it will be necessary to remove the pivot pin and fit a .040" (1 m/m) thick packing washer ('C' Fig. H15), Part Number R.4468, between the oil trough 'E' and the inner roller race.

(c) If dimension 'D' is over .515" (13 m/m), fit two packing washers.

(iii) Having refitted the pivot pin and fully tightened nut, check that the stub axle can be moved freely from one full lock to another.

(iv) Continue the re-assembling operatings.

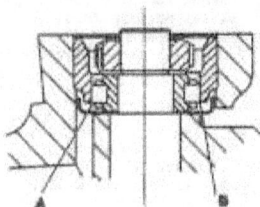

FIG. H14.

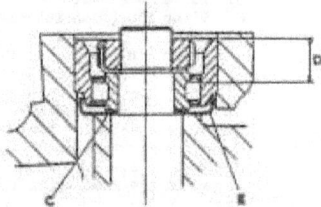

FIG. H15.

### To Refit Stub Axle to Yoke:

(i) Ascertain that the dowel screw fitted to the yoke is tight.

(ii) Early type assembly, replace oil trough (17), with lip uppermost. Later type assembly, replace felt washer (28) distance piece (27). Place the metal packing ring and felt washer sub-assembly (16 and 21) in the recess provided, i.e. with the packing ring against the yoke. Renew the felt washer if necessary and soak in oil before fitting.

(iii) With the 34 needle rollers in position in the lower bearing, place the stub axle on to the yoke. Enter the pivot pin as far as it will go by hand, sighting upon entering to make sure that the narrow slot in the top of the pivot pin is lying in a radial line with the dowel screw (3, Fig. H16). The pin must be driven in until the shoulder on the pin makes contact with the corresponding shoulder in the yoke.

(iv) Fit new rollers and cage assembly into position on the pivot pin. Tighten up the nut (X.24), using a box spanner, but do not bend up the tabs of the lockwasher at this stage.

(v) Lubricate and place the thrust washer (25) into the recess in the bottom cover. The thickness of a new thrust washer is .098" - .001" and the depth of the recess in the bottom cover is .100" + .004". The fitting allowance (end lift) of a stub axle is from .007" to .017".

SILVER WRAITH — SILVER DAWN — BENTLEY MK. VI.

R. TYPE BENTLEY — PHANTOM IV.

(vi)    Smear the joint face of the bottom cover with a jointing compound and
        fit it to the stub axle fully tightening the nuts. Ensure that the
        stub axle can be freely moved from one full lock to another. If found
        to be stiff, then with a soft aluminium drift, give the lower cover a
        few sharp taps upwards and CHECK THE PIVOT PIN NUT (X.24) FOR TIGHTNESS.

(vii)   Place the restricting pin into the pivot pin, lubricate the upper
        bearing with oil. Smear the joint face of the top cover (8) or (9)
        with a jointing compound and fit the cover. Fit oil feed (tube 18)
        early type, to the bottom cover.

(viii)  On completion of assembly, and having reconnected the brake pipe,
        bleed the front braking system.

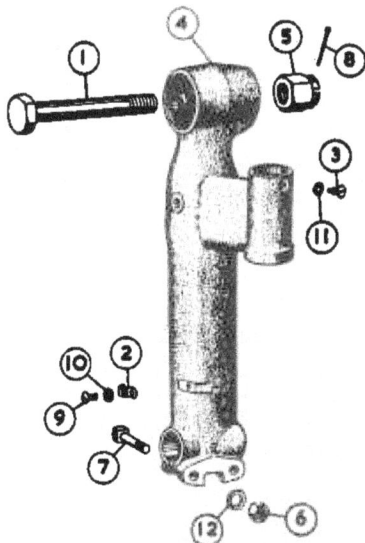

FIG. H16.   THE YOKE TO LOWER TRIANGLE LEVER.

| | |
|---|---|
| 1. Bolt (Upper Bearing). | 7. Bolt (Split Boss). |
| 2. Clip (Oil Pipe). | 8. Split Pin (Upper Bearing). |
| 3. Grub Screw (Locating). | 9. Screw (Oil Pipe Clip). |
| 4. Yoke. | 10. Washer, spring (Oil Pipe Clip). |
| 5. Nut (Upper Bearing). | 11. Washer, Spring (Grub Screw). |
| 6. Nut (Split Boss). | 12. Washer, Spring (Split Boss). |

SILVER WRAITH — SILVER DAWN — BENTLEY MK. VI.

R. TYPE BENTLEY — PHANTOM IV.

### TO REMOVE YOKE FROM LOWER AND UPPER TRIANGLE LEVERS:

(i)     Remove the nut and spring washer (6, Fig. H17), the cover (5) and the plain washer (7) from the bolt (18). Disconnect the coiled oil pipe (4) (inside cover) from the three-way junction on the torque arm and elbow connection fitted to the front of the yoke. Remove the clip from the yoke, then pipe and distance piece (8).

(ii)    Remove the split pin, castellated nut (9), and aluminium washer (10). Remove the two clamping bolts from the lower end of yoke. Tap out bolt (18). Remove roller housing (3), together with the adjusting washer (2). Remove bearing pin and distance piece (21 and 13) which will remove with it the rear roller housing and adjusting washer (17 and 20) from the lower triangle lever. Collect the needle rollers, remove roller retaining washer (15 and 16) and felt washer (1 and 14) from the lever.

(iii)   Remove the split pin, nut and bolt from the Silentbloc bearing at the upper end of the yoke and remove the yoke.

### Silentbloc Bearing:

If the rubber of the Silentbloc bearing at the upper end of the yoke has collapsed, a new bearing must be fitted with the aid of a press. If the needle roller path of the roller housings (3 and 17) and the ends of the bearing pin (21) show load markings from the needle rollers, new parts should be fitted.

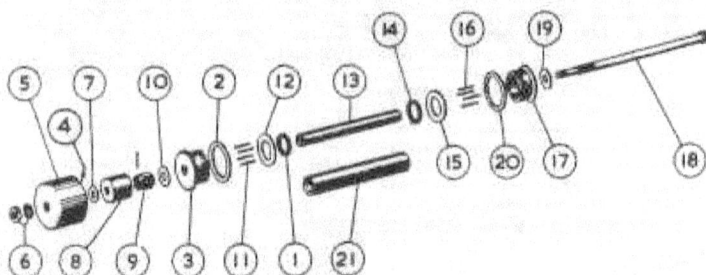

FIG. H17.  YOKE LOWER BEARING.

| | |
|---|---|
| 1. Felt Washer. | 12. Washer, Roller retaining. |
| 2. Adjusting Washer (Range of). | 13. Distance Piece. |
| 3. Roller Housing. | 14. Felt Washer. |
| 4. Coiled Oil Pipe (Inside Cover). | 15. Washer, Roller Retaining. |
| 5. Cover - Coiled Oil Pipe. | 16. Needle Roller (27 off) 2.5 x |
| 6. Nut and Spring Washer. | 15.8 mm). |
| 7. Plain Washer. | 17. Roller Housing. |
| 8. Distance Piece (Assembly). | 18. Bolt. |
| 9. Nut, Castellated. | 19. Washer, aluminium. |
| 10. Washer, aluminium. | 20. Adjusting Washer (Range of). |
| 11. Needle Roller (27 off) 2.5 x | 21. Bearing pin. |
| 15.8 mm). | |

TO ASSEMBLE LOWER BEARING OF YOKE AND ADJUST END FLOAT OF ROLLER HOUSINGS:

(i)    Remove the existing adjusting washers (2 and 20, Fig. H17). Fit the roller housing (3), the bolt (18), the distance piece (8), and the roller housing (3), to the lower triangle lever and tighten up the nut (9). Do not connect the yoke to the lower bearing (triangle lever).

(ii)    Determine the thickest adjusting washers (2 and 20) which can be pushed in between the lower triangle lever and the roller housings. These should be of equal or adjacent thickness, i.e. the difference in thickness between the two washers should not exceed .005". This should give .000" to .005" end float of the roller housings and distance piece in the triangle lever. Remove the bolt, the roller housings and the distance piece.

(iii)    Fit the bearing pin (21) to the lever and yoke. Place the felt washers (1 and 14) and the roller retaining washers (12 and 15) in position in the lever with chamfered side of washers towards the felt washers. Fit the distance piece. Place the adjusting washers selected on the roller housings with chamfered side against flange. Smear the inside of the roller housings with grease, using the minimum amount to hold rollers in position. Fit the needle rollers in each housing. Ensure that the oil holes in the housings are not blocked with grease. Oil the rollers and fit the assemblies into the lever. Fit the bolt (18), using new aluminium washers (10 and 19), tighten up nut (9) and lock.

(iv)    Centralise the yoke in the fork of the lower triangle lever as far as the two slots in the bearing pin (21) will allow, and refit the bolts. Refit the bolt to the upper triangle lever and yoke, with the head of the bolt to the front of the chassis. Screw up castellated nut and temporarily secure with a split pin. This nut must be tightened up and locked when front suspension is in normal ride position, i.e. with passengers in front seats and road wheels on the ground. Fit the distance piece (8, Fig. H17) on to the bolt, the coiled oil pipe (4) over the distance piece, and attach the pipe to the yoke. Reconnect the pipe to elbow. Fit plain washer (10), cover (5), and secure with spring washer and nut (6). Check that oil pipe does not foul the cover. With passengers in front seats, <u>tighten the castellated nut at the upper end of the yokes and secure</u>.

FRONT SPRINGS:

On early series chassis, the springs were .600" shorter than current productions, the rating is the same. Bentley export models may be fitted with "Colonial" type or a heavier rated type, of springs to customers requirements.

Packing washers .062" thick, up to ten in number may be used to raise the "standing height". The fitting of one washer raises the standing height by ¼". If a number of packing washers are fitted, these should be divided and fitted at the top and bottom of spring.

REMOVAL AND REPLACEMENT OF FRONT SUSPENSION SPRINGS:

To Remove:

(i)    Remove the castle nut securing the ball end pin to the side steering lever. Disconnect the outer end of the cross steering tube from the cross steering lever.

With a steady block against the lever at the ball and pin end, tap the eye of the lever out with a drift. Remove nuts and spring washers from the bracket of the front stabiliser and lift from studs. Place the spring retaining bolt tool No. 3752/T1002 downwards through the hole (A, Fig. H18) provided in the top of the front pan and leave in this position.

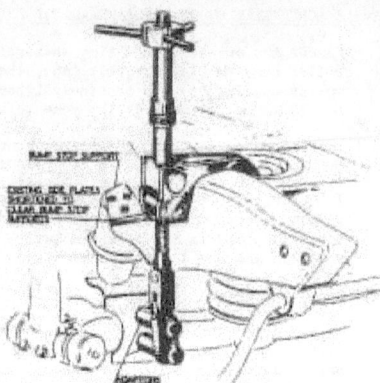

FIG. H18. SPRING COMPRESSION TOOL IN POSITION.

(ii) Place the lower end of the special tool No. 3752/T1008 on the studs of the front stabiliser bracket, with claws fitting under the lip (edge) of the hole in the top of the front pan. Tighten knurled nuts to secure. Compress the spring then fully tighten spring retaining bolt. While compressing spring, guide threaded end of retaining bolt through the hole in the lower buffer stop leaving the spring in position (Fig.H19).

(iii) With the hub held, remove the bolt from the Silentbloc bearing at the upper end of the yoke, and allow the hub to rest on a wooden block. Release the compression tool, and allow the triangle lever to move downwards until the rubber buffer attached to the upper triangle lever rests on its stop. Remove the spring complete with bolt and temporarily reconnect the yoke to the upper triangle lever by means of a tommy bar.

To Replace:

(i) To fit a spring, a "Pot" Special Tool STD-416 will be required to compress and decompress the road spring. Place the spring in the "Pot" and clamp it down fully by tightening the two outer nuts of the "Pot". Remove the retaining bolt from the spring.

(ii) With one of the special washers on the bolt, pass the "Pot" centre bolt through the bottom of the "Pot", place the other washer on the cover, and fully tighten the bolt. Remove the outer "Pot" nuts, then gently unscrew the nut on the centre bolt and allow the spring to expand to its full free length.

(iii) With the fabric seatings (1 and 5, Fig. H19), bump stop and adjusting washers (if required), in position, fully compress the spring. Remove the centre bolt from the "Pot", and in its place fit the retaining bolt, and fully tighten the nut.

(iv) Fit the spring to the vehicle. With the spring in position and fully compressed by the compression tool, the yoke should be reconnected to the upper triangle. This must be done before removing the spring retaining bolt. The castellated nut of the yoke bolt may only be fully tightened and split pinned when the car has been lowered to the ground

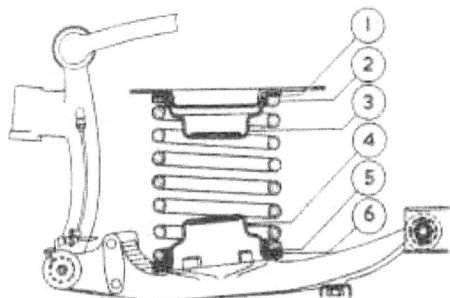

FIG. H19. FRONT SUSPENSION SPRING IN POSITION.

1. Spring Seating (Rubberised fabric).
2. Road Spring.
3. Buffer Stop (Upper).
4. Buffer Stop (Lower).
5. Spring Seating (Rubberised fabric).
6. Adjusting Washer, if required.

and bumped a few times, to settle to its normal loaded state. Should the spring touch the frame, it should be compressed and rotated to a suitable position.

## TO REMOVE AND REFIT LOWER TRIANGLE LEVERS:

(i) Remove the front springs and with the yokes temporarily reconnected to the upper triangle levers of the front shock dampers, remove the two stays connected to the lower triangle levers and torque arms.

(ii) Remove the oil feed pipe from the elbow connection on each of the lower levers and disconnect the yokes from the lower levers.

(iii) Remove nuts and setscrews from centre (jacking) plate, do NOT remove the two ¾" B.S.F. nuts near the centre. Remove the four 5/16" set-screws, two at the front of the plate and two at the rear, only visible from the upper side. Lower the centre plate from the front pan, exposing the bracket to which the inner ends of the lower triangle levers are attached. Remove the levers from the centre bracket and the bracket from the plate. Check centre bracket for trueness, the principal dimensions are shown in Fig. H20. Replace Silentbloc bearings if necessary.

(iv) Refit the lower triangle levers and check outer ends for correct height as follows:-

(a) Refit bracket to centre plate, inserting setscrews and bolts from the top. Assemble the triangle levers to the bracket, fitting the two bolts so that the castellated nuts face to front of car and lightly tighten the nuts.

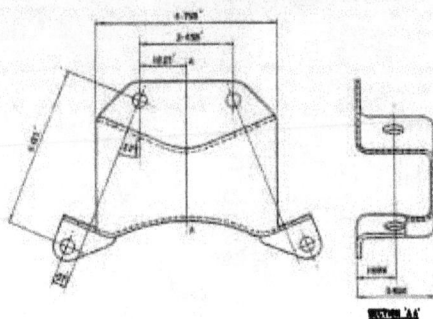

FIG. H20.   CENTRE BRACKET - LOWER TRIANGLE LEVERS.

(b)  Place a long straight edge against the underside of the centre
     plate and move one of the triangle levers until the centre of
     the bore in the outer end of the lever is .200" below the under-
     side of the centre plate as shown in Fig. H21. Tighten up the
     castellated nut and secure. While tightening the nut, check that
     the lever has not moved from the .200" position.

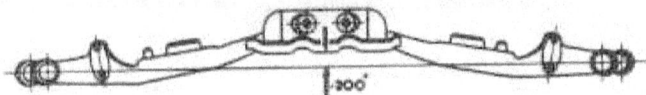

FIG. H21.

### TO FIT NEW CENTRE (JACKING) PLATE TO FRONT PAN:

The centre plate or jacking bracket is drilled to register with the holes
in the centre pan, above it. The two tapping strips, however, must be marked off,
drilled, and tapped as required, using the pan as a template.

### TO FIT NEW RUBBER SPHERICAL BEARING BUSH TO REAR END OF TORQUE ARM:

If the rear end of a torque arm can be moved sideways by hand, the
rubber bearing has collapsed.

(i)   Remove the nuts and spring washers retaining the cap to the bracket and
      pull the rear of the torque arm downwards. Remove the rubber bearing
      bush. The spherical end of the torque arm should be cleaned up and
      polished with medium and fine emery cloth.

(ii)  Press the new bush on to the torque arm with the two locating "pips"
      uppermost. Prior to fitting the bush, squeeze out as much air as
      possible from it, as soon as the mouth of the bush touches the ball
      end of the torque arm and rotate while pressing it on.

- H.29 -

(iii)   Push the end of the torque arm into position against the bracket
ensuring that the two "pips" of the bush have entered the corresponding
holes in the bracket.

(iv)   Fit the cap, ensuring that the lower "pip" of the bush has engaged
with the corresponding hole in the cap, then tighten the nuts.  Slacken
off the nuts about $1\frac{1}{2}$ turns, bounce the car at the front and finally
tighten up the nuts.

SILVER WRAITH — SILVER DAWN — BENTLEY MK. VI.

R. TYPE BENTLEY — PHANTOM IV.

## FRONT SHOCK DAMPERS.

### Loading

| | | |
|---|---|---|
| Up (Bump) | - | 65 - 75 lbs. |
| Down (Rebound) | - | 125 -135 lbs. |
| Oil | - | SAE.20. |

### GENERAL.

The shock dampers are of the double acting type and consist of two pistons operating in oil filled cylinders, the oil being displaced from one cylinder to the other through drilled passages. The degree of damping being controlled by spring loaded valves. Recuperating valves are fitted in bottom of each piston.

### REMOVAL.

(i)   Jack up car and place blocks under the outer ends of lower triangle levers.

      NOTE:- Weight of car must not be taken off blocks until damper is refitted or front spring will be displaced.

(ii)  Remove wheel, and disconnect upper triangle lever from stub axle yoke, support hub on wooden block. Remove damper from frame, collect distance washers.

### DISMANTLING.

(i)   Remove top cover. Unscrew valve caps (13 and 27, Fig.H.22), about half way and drain oil. Remove rebound valve assembly, situated under detachable lever. Remove bump valve assembly from other side. Keep parts separate.

(ii)  Remove distance piece carrying rubber rebound stop (1). Slacken pinch bolt and remove lever (2).

(iii) Slacken off pinch bolt on rocker (8). Mark rocker and casing to ensure that pistons are refitted in original bores.

      To ensure that splines of main shaft are engaged with same splines on rocker on re-assembly:-

      (a) Hold lever in midway position so that gap in top of rocker is in T.D.C. position, scribe a line across the boss of lever and adjacent boss of main casing in line with gap of rocker.

SILVER WRAITH — SILVER DAWN — BENTLEY MK. VI.

R. TYPE BENTLEY — PHANTOM IV.

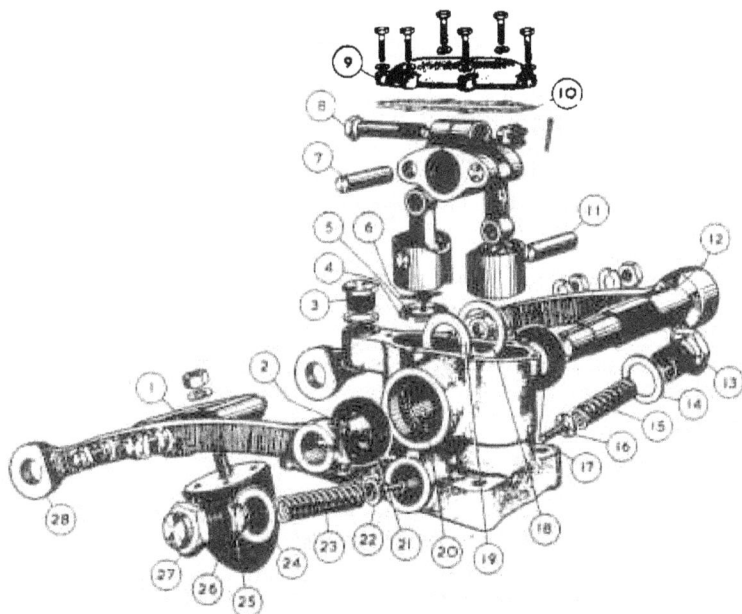

FIG. H.22. FRONT SHOCK DAMPER.

1. Buffer support.
2. Pinch bolt.
3. Filler plug.
4. Spring ring.
5. Dished plate.
6. Replenishing valve assembly.
7. Pin.
8. Rocker bolt.
9. Cover.
10. Joint washer.
11. Pin.
12. Upper triangle lever and mainshaft assembly.
13. Valve cap.
14. Aluminium washer.

15. Bump valve spring.
16. Bump valve.
17. Large gland rubber.
18. Large bearing washer.
19. Small bearing washer.
20. Small bearing bush.
21. Small gland rubber.
22. Rebound valve.
23. Rebound valve spring.
24. Aluminium washer.
25. Adjusting washer (When fitted).
26. Buffer.
27. Valve cap.
28. Upper triangle lever.

(b) Hold lever hard up and scribe a second line on boss of lever in line with previous line.

(c) Hold lever hard down and scribe a third line on boss. Tap out main shaft and lever.

(iv) Remove rocker and piston, bronze thrust washers (18 & 19) and rubber gland (17 & 21). Remove filler cap (3). Remove spring rings (4), dished plates (5) and replenishing valves (6).

## RE-ASSEMBLING.

Re-assemble damper in reverse order of dismantling. The bronze thrust washers should be renewed if they are scored. If the pins (7 & 11) are worn, replace.

Fill damper with oil and bleed by pumping levers until all free movement is lost.

## PISTON FIT IN BORES OF MAIN CASING.

Selective fitting of pistons is carried out to reduce working clearances to a minimum. Both piston and bores are colour coded and fitted colour to colour. The piston is coloured on its pin boss and the bore at its base:-

| PISTON DIAMETER. | BORE DIAMETER. | COLOUR. |
|---|---|---|
| 1.449 - 1.4495 | 1.500 - 1.5005 | Red |
| 1.4459 - 1.500 | 1.5005 - 1.501 | Green |
| 1.500 - 1.5005 | 1.501 - 1.5015 | Blue |
| 1.5005 - 1.501 | 1.5015 - 1.502 | Yellow |

## POUNDAGE ADJUSTMENT.

If shock damper test rig is available, adjustment to poundage can be made by varying size of adjusting washers in valve caps.

In certain countries where additional damping may be required, it is permissible to plug one of the two oil leak holes provided in each valve by removing either the bump or rebound valve and soldering up the leak hole in the head of the valve which communicates with the leak hole in the stem of the valve near the head. These two holes should not be confused with the three larger diameter equally spaced holes drilled near the centre of the valve stem.

# SECTION
# J
# REAR AXLE AND SPRINGS

SILVER WRAITH — SILVER DAWN — BENTLEY MK. VI.

R. TYPE BENTLEY — PHANTOM IV.

SECTION J

## REAR AXLE AND SPRINGS

REAR AXLE SHAFTS - BEVEL PINION AND CROWN WHEEL -
DIFFERENTIAL GEARS - REAR ROAD SPRINGS - REAR SHOCK
DAMPER.

SILVER WRAITH — SILVER DAWN — BENTLEY MK. VI.

R. TYPE BENTLEY — PHANTOM IV.

SECTION J

REAR AXLE AND SPRINGS

List of Illustrations:

SILVER WRAITH — SILVER DAWN — BENTLEY MK. VI.

R. TYPE BENTLEY — PHANTOM IV.


REAR AXLE AND SPRINGS


Rear Axle Ratios:  11:41 Standard Silver Wraith, Silver Dawn, Bentley.
12:41 Special export Bentley.
13:40 Bentley Continental Sports.
8:34 Silver Wraith (Long Wheelbase), Phantom IV.

Oil Capacity:  1½ pts. Silver Wraith (Standard), Silver Dawn, Bentley.
3 pts. Silver Wraith (Long Wheelbase), Phantom IV.


REAR AXLE:

General:

The rear axle is of the semi-floating type with off-set hypoid spiral bevel final drive. The load is taken by single row journal ball bearings at the outer ends of the axle shafts, each bearing being permanently lubricated and sealed.

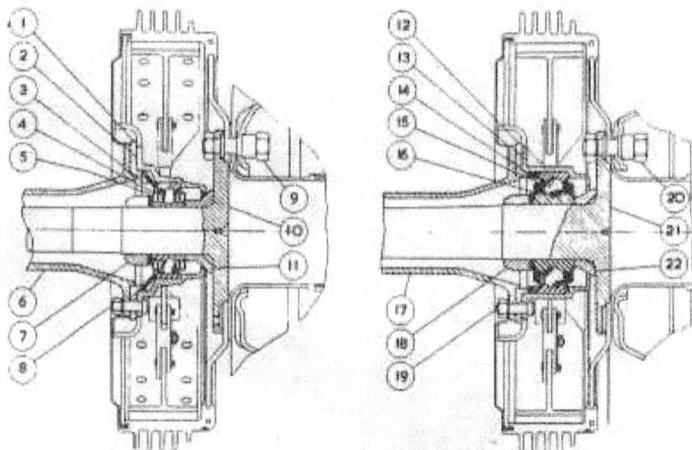

EARLY TYPE                FIG. J1.                LATER TYPE
AXLE BEARING ASSEMBLIES

| 1 & 12. | Spigot plate | 6 & 17. | Axle tube |
|---|---|---|---|
| 2 & 13. | Bearing housing | 7 & 18. | Collar |
| 3 & 14. | Adjusting piece | 8 & 19. | Bolt |
| 4 & 15. | Ball bearing | 9 & 20. | Nut, wheel studs |
| 5 & 16. | Spring plate | 10 & 21. | Axle shaft |
| | | 11 & 22. | Distance piece |

- J.1 -

SILVER WRAITH — SILVER DAWN — BENTLEY MK. VI.

R. TYPE BENTLEY — PHANTOM IV.

The axle shafts are forged integrally with the wheel hubs and can be extracted without dismantling the axle.

Axle Shafts:

To remove a half-shaft:-

1. Remove wheel, undo the three countersunk brake drum retaining screws, withdraw wheel disc support and drum.

2. Remove the five $\frac{3}{8}$" bolts, Phantom IV ten setscrews, retaining bearing housing to axle tube, hold brake carrier in position and withdraw shaft. The shaft should be withdrawn carefully to avoid damage to oil seals adjacent to crown wheel bearing.

It should be noted, that the axle shafts differ in length, the R.H. shaft being longer than the L.H. shaft.

On early models, the wheel bearing housings 2, Fig. J1, are different to current production, 13, Fig. J1, the axle shafts also being of a smaller diameter. They are interchangeable as a whole and the procedure for dismantling and assembly are similar.

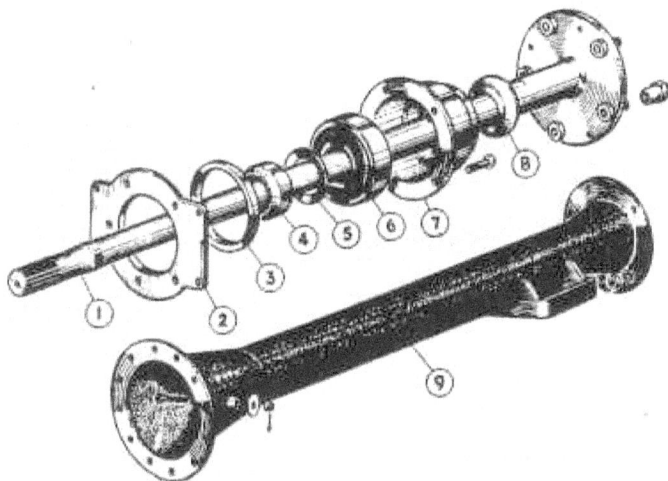

FIG. J2.    AXLE ASSEMBLY.

1. Axle shaft.
2. Spigot plate.
3. Adjusting piece.
4. Collar.
5. Spring plate.
6. Ball bearing.
7. Housing.
8. Distance piece.
9. Axle tube.

SILVER WRAITH — SILVER DAWN — BENTLEY MK. VI.

R. TYPE BENTLEY — PHANTOM IV.

Data, Half-Shaft Assembly

Bentley - A & B series:

| | |
|---|---|
| Interference fit, bearing and shaft | .0006" to .0014" |
| Interference fit, collar and shaft | .003" to .0043" |

Bentley - (C series and onwards), Silver Wraith, Silver Dawn:

| | |
|---|---|
| Interference fit, bearing and shaft | .0006" to .0014" |
| Interference fit, collar and shaft | .0038" to .0046" |

Phantom IV:

| | |
|---|---|
| Interference fit, collar and shaft | .0010" |

Axle bearing end float (All)  .010" to .015"

Attention is drawn to the appreciable end float in the above bearings and they should not be renewed unless the end float considerably exceeds this figure or they are rough in operation.

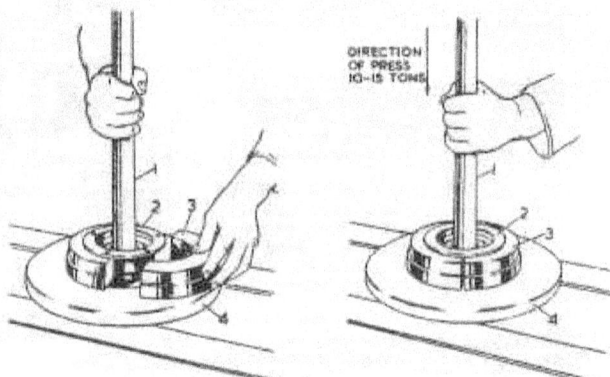

FIG. J3.  REMOVING BEARING AND HOUSING.

1. Axle shaft
2. Bearing and housing
3. Split collar
4. Support plate

To Renew Bearings:

A special extractor 733/T1002 is required.

1. Check studs for alignment. With the spigot plate (1 & 12, Fig. J1) removed from the bearing housing, mount assembly in lathe, turn the collar (7) until thin enough to split. Check shaft for bowing, if more than .002", straighten in cold press.

# SERVICE HANDBOOK

### SILVER WRAITH — SILVER DAWN — BENTLEY MK. VI.
### R. TYPE BENTLEY — PHANTOM IV.

2. Use extractor, or remove bearing as shown in Fig. J3.
NOTE:- If shaft is worn, new diameter 1.7726" - .0003", due to
bearing revolving on shaft, a complete new assembly must be fitted.
If the oil seal has worn a small groove, blend edges by filing,
providing this is not too deep.

3. Measure the width of the outer race of the new and old bearing, the
variation should not exceed .002". If the new bearing is wider, a
similar amount should be ground off the existing adjusting piece
(3 and 14, Fig. J1), but if it is narrower or of the same width, the
adjusting piece may be fitted without alteration. A .002" end float
of adjusting piece is permissible.

3a. If a new bearing housing and spigot plate
are to be fitted, press the new bearing
fully home into housing and bolt on spigot
plate without fitting adjusting piece.

Place assembly on surface plate and clock
top face, as shown in Fig. J4. Fit new
adjusting piece and again take clock
reading.

An amount equal to the difference in
clock readings should then be ground
off the adjusting piece, to obtain
zero nip on the bearings, but a nip
not exceeding .002" is permissible.

4. Place the adjusting piece (3 or 14)
Fig. J1, onto the axle shaft followed
by the housing complete with bearing,
press bearing fully home (approx: 10
ton pressure). Fit the spring plate
(5 or 16) ensure this is correct way round.

FIG. J4.   CLOCKING TOP FACE
OF SPIGOT PLATE.

5. Press on collar (7 or 18) which will flatten out spring plate, place
the adjusting piece in position, attach spigot plate to housing with
two bolts to hold correct assembly.

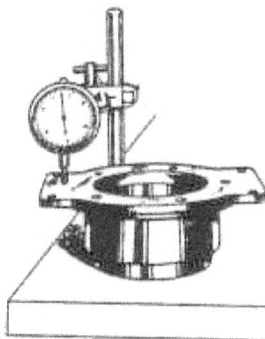

### To Remove Rear Axle:

1. Jack up and place blocks under rear shackle brackets. Disconnect
chassis lubrication unions.

2. Disconnect propeller shaft at rear end. Disconnect brake rods from
equaliser, and equaliser support from axle tube.

3. Jack up rear axle to free shock damper arms, and disconnect forked
links from main arms.

4. Disconnect the four U-bolts and remove near-side brake drum. Manoeuvre
axle casing out between springs and chassis frame towards the off-side.

### To Replace Rear Axle:

Replacement is the approximate reversal of the removal operations.

Renew rubber mounting pads if necessary.

### SILVER WRAITH — SILVER DAWN — BENTLEY MK. VI.
### R. TYPE BENTLEY — PHANTOM IV.

Check rear equaliser support is correctly positioned, see Fig. J5.

DISMANTLING CENTRE CASING:

Before dismantling, suitably mark with co-relation marks axle tubes and centre casing, also relative positions of side plates and centre casing.

Remove axle tubes from centre casing complete with shafts and brake drums.

Remove side plates, place casing under press and remove nuts progressively in opposing pairs to avoid distortion from internal pre-load spring. Remove oil seal and adjusting washer (3 and 5, Fig. J6).

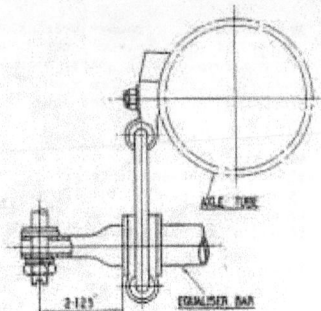

FIG. J5.   REAR BRAKE EQUALISER.

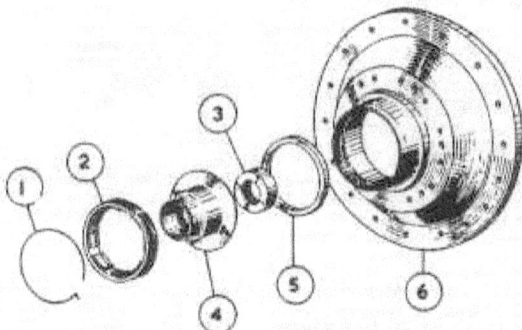

FIG. J6.   L.H. SIDE PLATE ASSEMBLY.

1. Locking ring.
2. Internally serrated nut.
3. Oil seal.
4. Oil seal housing.
5. Adjusting ring.
6. Side plate.

Lift out crown wheel assembly from L.H. side.

Remove nuts from pinion retaining plate and withdraw bevel pinion assembly. Remove spring housing (9, Fig. J7) with press or clamp.

BEVEL PINION:

From Bentley chassis B.433.SP (12/41 axle) B.380.SR (11/41 axle), Silver Wraith chassis WVH.77, Silver Dawn chassis SMF.42, taper roller bearings are fitted in place of the multiple bearing for the pinion, see Figs. J8 and J9.

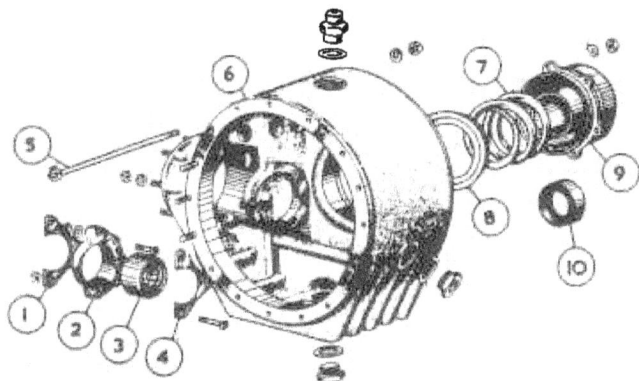

FIG. J7.  CENTRE CASING - BENTLEY, SILVER WRAITH, SILVER DAWN.

1 & 4.  Bridge plates, nose bearing.
2.  Bridge, pinion nose bearing.
3.  Bearing.
5.  Retaining bolt
6.  Casing.

7.  Spring.
8.  Thrust washer.
9.  Spring housing.
10.  Oil seal.

## To Renew Multiple Bearing:

1.  Place nose of pinion in suitable holding block in vice.  With Tool No. 1649/T3, remove driving flange retaining nut and lock-washer and withdraw flange with suitable extractor, collect Woodruff Keys. Remove bearing housing cover.

2.  With spanner No. 1649/T2, remove pinion bearing retaining nut (left-hand thread).  Press out pinion. Press out bearing from housing.

3.  Select adjusting washer (4, Fig. J8) to give .003" gap between cover and housing when cover is pressed down firmly by hand.

4.  Lubricate and re-assemble. Fit new oil seal felt in bearing housing cover and replace driving flange.

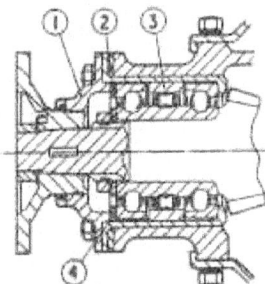

FIG. J8.  PINION AND MULTIPLE BEARING.

1.  Cover.
2.  Pinion adjusting washer.
3.  Bearing.
4.  Bearing adjusting washer.

1. Oil seal.
2. Flinger.
3. Bearing housing.
4. Outer race.
5. Adjusting washers.
6. Outer race.
7. Pinion depth adjusting washer.
8. Pinion bearing retaining nut.
9. Flange retaining nut.

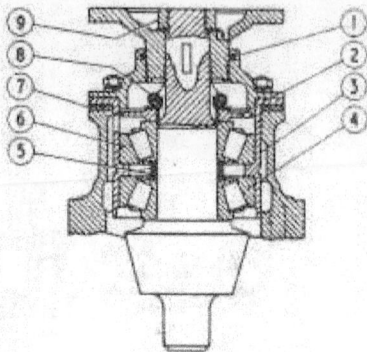

FIG. J9.   PINION AND ROLLER BEARING.

## To Renew Taper Roller Bearing:

The correct pre-load on these bearings is assessed by the drag torque measured at the holes in the pinion bearing housing flange and is adjusted by selective fitting of adjusting washers between the inner races. The drag torque should be 5-12 in./lbs. with bearings dry.

The London Service Station will undertake the fitting and pre-loading of new taper roller bearings if the bearing housing and old adjusting washers are returned to them.

1. Place nose of pinion in suitable holding block in vice. With Tool No. 1649/T3, remove driving flange retaining nut and lockwasher and withdraw flange with suitable extractor, collect keys. Remove bearing housing cover.

2. With spanner No. 1649/T2 remove pinion bearing retaining nut (left-hand thread). Press out pinion.

3. Collect outer bearing and adjusting washers. Remove inner bearing from pinion shaft with suitable extractor. Tap out outer races from housing.

4. Tap new outer races into bearing housing. Assemble new bearings and housing, dry, on dummy pinion of Gauging Tackle as shown in Fig. J10.

   Tighten knurled nut until drag torque is between 5 and 12 in/lbs.

5. With micrometer, measure overall dimension between outer face of flange of pinion and outer face of knurled nut as shown in Fig. J10. Record dimension.

   Strip assembly and measure overall width of inner race. The total width of the two inner races plus 2.00", deducted from recorded dimension will give width of adjusting washer for correct pre-load.

Select two adjusting washers to give this thickness and re-assemble pinion and housing, 5 to 12 in/lbs., overall thickness .352" to .378".

If special tool STD.717 is not available, pre-loading can be carried out by trial and error, using mandrel as shown in Fig. J11, on which the inner races of the bearing are a sliding fit, and an accurate spring balance clipped to one of the holes in the flange of the pinion bearing housing.

1. Tap new outer races into bearing housing and assemble in dry conditions on mandrel held vertically in vice, using the two original adjusting washers. See Fig. J12.

2. Fit L.H. thread retaining nut and whilst gradually screwing down, turn housing by hand to ensure that no undue load is applied to bearings. Thicker adjusting washers must be fitted if drag, measured by spring balance begins to exceed 6 lbs. Do not crush bearings.

3. Correct pre-load is with nut fully tightened and ..... balance shows a drag of between 2½ and 6 lbs.

4. When assembling bearings to actual pinion, increase thickness of adjusting washers by .002" to allow for expansion of inner races when pressed onto pinion. This expansion is allowed for when special tool is used.

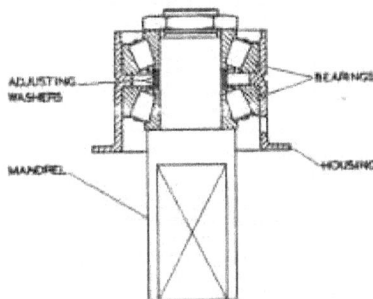

MICROMETER MEASUREMENT - 2.00 = WIDTH OF SPACER SHIMS REQUIRED.

FIG. J10.   GAUGING BEARING PRE-LOAD.

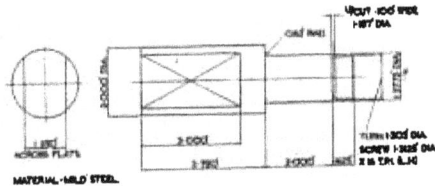

FIG. J11.   MANDREL - DETAILS FOR MANUFACTURE.

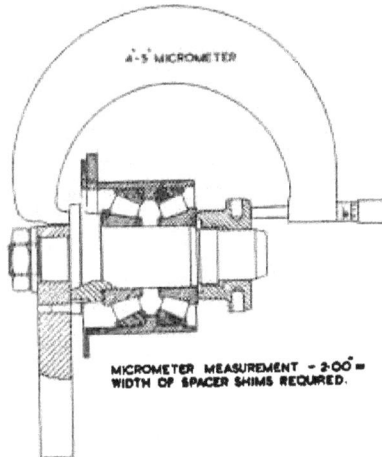

FIG. J12.   BEARING ASSEMBLY MOUNTED ON MANDREL.

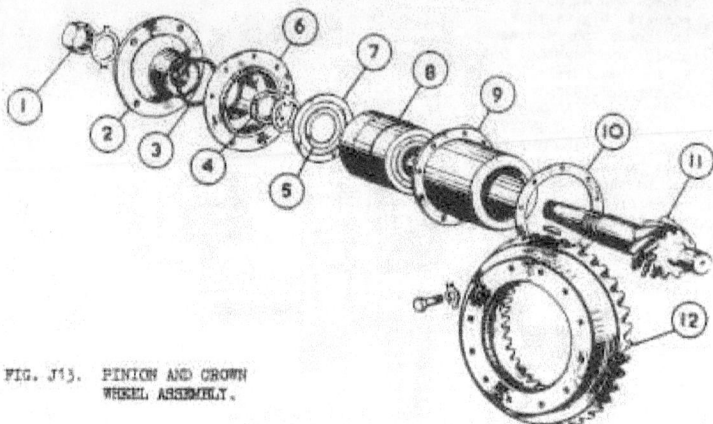

FIG. J13. PINION AND CROWN
WHEEL ASSEMBLY.

1. Nut, driving flange.
2. Driving flange.
3. Felt washer.
4. Nut, pinion bearing.
5. Adjusting washer.
6. Cover.

7. Adjusting washer.
8. Bearing.
9. Bearing housing.
10. Adjusting washer.
11. Bevel pinion.
12. Crown wheel.

Re-assemble and refit to axle casing.
Bearings should be light press fit .00025" - .0005" on shaft.

CROWN WHEEL AND PINION:

The crown wheel and pinion are
supplied in lapped pairs and must not be
used independently.

Crown wheel and pinion may be
stoned if necessary, but they must be
reset to original figures, etched on
parts, by means of new adjusting washers
if required.

To Check Crown Wheel:

Place crown wheel complete with
roller bearings and outer races under a
press with suitable distance pieces "A"
and "B" as in Fig. J14.

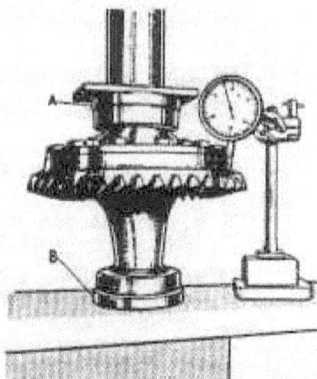

FIG. J14. CHECKING CROWN WHEEL RUN-OUT.

# Service Handbook

### SILVER WRAITH — SILVER DAWN — BENTLEY ᴹᵛ ᵛᴵ
### R. TYPE BENTLEY — PHANTOM IV.

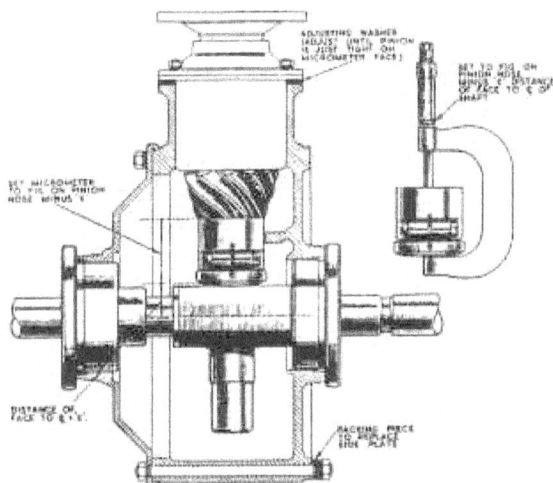

FIG. J15. PINION SETTING TOOL IN POSITION.

Apply light pressure and rotate wheel slowly. The "run-out" should not exceed .002". If over this figure, inspect crown wheel for incorrect seating on differential casing.

### To Fit New Crown Wheel and Pinion:

1. Fit replacement pinion into pinion housing, refit driving flange but do not lock.

2. Set pinion depth in relation to crown wheel:-

   a) Subtract 1" from the dimension etched on pinion behind nose bearing.

   b) Set measuring piece, see Fig. J15, to resultant dimension and lock.

   c) Fit L.H. side plate to centre casing with packing pieces under bolt heads and fully tighten.

   d) Fit pinion setting tool into casing, see Fig. J15, using original adjusting washer, chamfer upwards, under pinion housing flange.

   e) Tighten down gradually, using three equally spaced nuts. The correct pinion depth is with no end float at measuring piece and with nuts fully tightened. The pinion will be slightly stiff due to nip on bearing, check actual clearance by moving measuring piece.

### SILVER WRAITH — SILVER DAWN — BENTLEY MK. VI.
### R. TYPE BENTLEY — PHANTOM IV.

If the original spacing washer (2, Fig. J8, or 7, Fig. J9) is not of suitable thickness, select one from range.

Fit remaining nuts, tighten down and re-check. Remove tool.

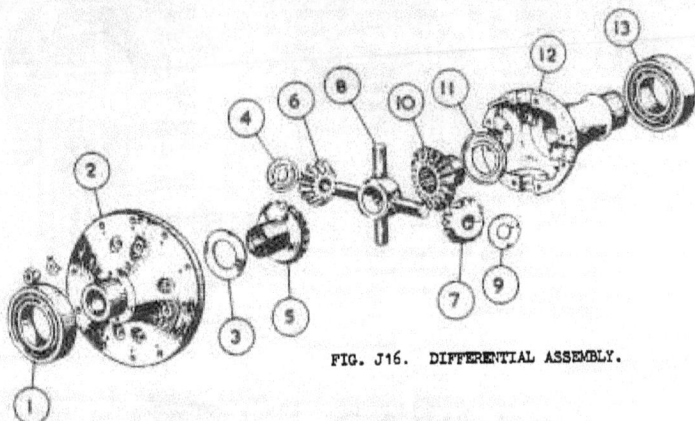

FIG. J16.  DIFFERENTIAL ASSEMBLY.

| | |
|---|---|
| 1. Bearing. | 8. Trunnion. |
| 2. Differential casing. | 9. Bevel adjusting washer. |
| 3. Bevel adjusting washer. | 10. Bevel, splined. |
| 4. Bevel adjusting washer. | 11. Bevel adjusting washer. |
| 5. Bevel, splined. | 12. Differential casing. |
| 6. Bevel, plain. | 13. Bearing. |
| 7. Bevel, plain. | |

NOTE:- The pinion setting tool Fig. J15, is not suitable for Silver Wraith Long Wheelbase or Phantom IV models.

When setting pinion on these models, the suggested method is by trial and error. The pinion teeth should be marked with blueing compound and the pinion and crown wheel assembled in the casing. Turn pinion by hand, dismantle and examine bedding. Adjustment can be made by varying the thickness of the adjusting washer as before.

3.  Fit new crown wheel to differential casing, do not lock.

4.  Fit the four bevel pinions on respective trunnion bearings and place assembly in R.H. half of differential casing without adjusting washers behind gears. Take Left-hand splined wheel and mesh with the four pinions. Draw the four pinions up so that the mitres at back all match correctly.

Measure gap between back of each bevel and casing and fit suitable washers, see Fig. J17.

- J.11 -

Mark relative positions and lift out trunnion assembly. Fit L.H. splined wheel with adjusting washer to give slight backlash. Place opposite splined wheel in casing with adjusting washer, bolt two halves together. Check backlash between splined wheels and bevels when fully tightened. Select washers to give zero backlash with whole assembly still free to turn. Dismantle, lubricate and re-assemble.

5. Fit new axle shaft seals.

6. Refit thrust washer, spring and housing to casing.

7. Note crown wheel backlash dimension, place crown wheel assembly with bearings in housing and refit side plates, no jointing compound.

8. Check and adjust pinion, crown wheel backlash:-

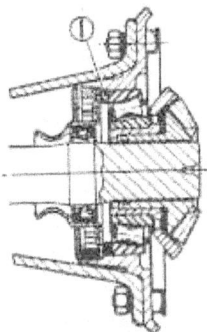

FIG. J17.  DIFFERENTIAL BEVEL ADJUSTING WASHER.

1.  Adjusting washer.

a) Tap L.H. outer race into position, select thinnest adjusting washer, grease and place in position. Fit oil seal housing and gradually tighten retaining nut at same time check that there is backlash between crown wheel and pinion.

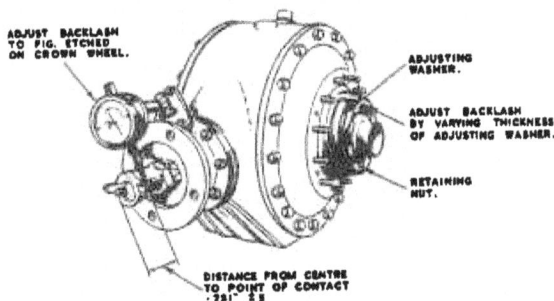

FIG. J18.  CHECKING PINION AND CROWN WHEEL BACKLASH.

b) Fit adaptor as shown in Fig. J18, take average figure of twelve readings. Adjust average reading to etched figure by varying adjusting washer above. Ratio of washer thickness to backlash is 1:1.

c) Complete assembly - lock.

9. Apply jointing compound to axle tube faces and refit tubes. Fit pinion driving flange.

SILVER WRAITH — SILVER DAWN — BENTLEY MK. VI.

R. TYPE BENTLEY — PHANTOM IV.

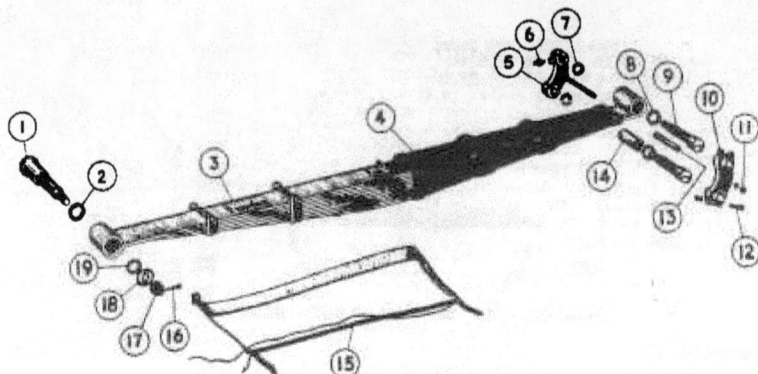

FIG. J19. REAR SPRING AND FITTINGS

1. Shackle pin, front.
2. Oil retaining washer, large.
3. Road spring.
4. Rear gaiter.
5. Inner shackle.
6. Oil connection.
7. Oil retaining washers, small.
8. Oil retaining washers, large.
9. Shackle pins.
10. Outer shackle.
11. Nut, shackle bolt.
12. Shackle bolt.
13. Distance piece.
14. Bush, rear shackle.
15. Front gaiter.
16. Split pin.
17. Nut, front anchor.
18. Washer.
19. Oil retaining washer, small.

## Oil Leak from Pinion Seal:

1.   Remove propeller shaft, mark before removal.

2.   Mark with co-relation marks pinion casing to differential casing,
     remove nuts and withdraw pinion.

3.   Remove nut on pinion, the oil thrower can then be removed to leave
     flange with felt oil seal.
     Fit new felt washers, refit flange using grease on felt and
     ensuring washer is not trapped.

## REAR SPRINGS:

The rear springs are of normal leaf type, adapted and strengthened
according to the terrain on which the car is to operate.

The forward ends of the rear springs are pivoted to the frame by means
of steel bushes. The shackle pins at the forward and rear ends are of the threaded
type, and both shackle pins and bushes are lubricated from the centralised chassis
system.

SILVER WRAITH  —  SILVER DAWN  —  BENTLEY MK. VI.

R. TYPE BENTLEY  —  PHANTOM IV.

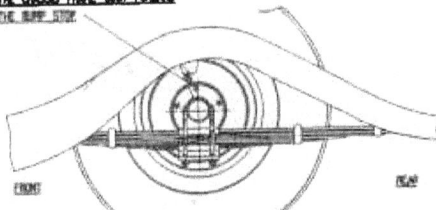

DIMENSION 'B' SHOULD BE TAKEN BETWEEN
THE AXLE TUBE & THE NEAREST POINT
ON THE CHASSIS FRAME JUST FORWARD
OF THE BUMP STOP.

FRONT                    REAR

FIG. J20.   REAR SPRINGS - STANDING HEIGHT.

Dimension "B":

| | | |
|---|---|---|
| Bentley Mk. VI | - | 5.625 |
| Silver Dawn | - | 5.625 |
| Silver Wraith | - | 5.765 |
| Phantom IV | - | 5.765 |

To Remove Rear Spring:

1.    Jack up and place support under chassis extremities.

2.    Remove U-bolt.  Disconnect lubricator pipe to shackle pin.

3.    Withdraw shackle pin.  Remove gaiters.

Spring Selection:

Spring selection should reproduce a figure for dimension "B",
Fig. J20, to give a nominal dimension as quoted, plus or minus .300", with
car at normal curb weight but free of passengers and driver.

If the petrol tank contains less than 5 gallons of petrol, .400"
should be added to this dimension.

On individual cars, the two dimensions on opposite sides of the
car should be within .378" of each other.

REAR SHOCK DAMPERS:

The shock dampers are of the double acting hydraulic type.  A
piston assembly operating in a cylinder maintained full of oil which is
displaced from one end of the cylinder to the other past a spring-loaded
valve.  The loading of this valve, hence degree of damping is controllable
though the "ride" control lever on the steering wheel.

The "Control" operates through a small gear type pump attached to
and driven from the gearbox to maintain a pressure of oil through piping to
each damper.  The pressure is variable, controlled by a spring loaded ball
valve, and is exerted on the valve in each damper through bellows.

SILVER WRAITH — SILVER DAWN — BENTLEY MK. VI.

R. TYPE BENTLEY — PHANTOM IV.

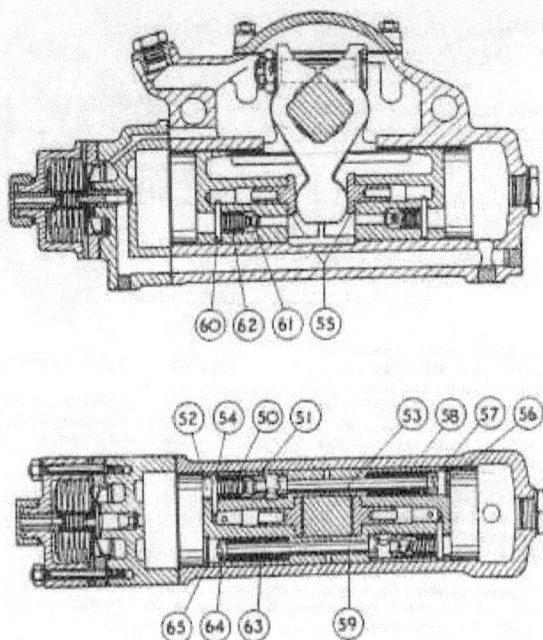

FIG. J21. SECTION - REAR SHOCK DAMPER

| | | |
|---|---|---|
| 50. Spring - Replenishing valve. | 58. | Spring - Piston bolt. |
| 51. Replenishing valve. | 59. | Bolt - Piston. |
| 52. Housing - Replenishing valve. | 60. | Taper pin. |
| 53. Bolt - Piston. | 61. | Replenishing valve. |
| 54. Taper pin. | 62. | Spring - Replenishing valve. |
| 55. Wearing pad. | 63. | Spring - Piston bolt. |
| 56. Expanding disc - Piston. | 64. | Nut - Piston bolt. |
| 57. Nut - Piston bolt. | 65. | Expanding disc - Piston. |

The pump draws its oil supply from the gearbox though this oil is not actually pumped into dampers. Check there is no wastage from this line.

A filter is provided in the gearbox for this oil, adjacent to oil level dipstick, and is removable for cleaning.

Dismantling Damper:

      To fit new gland packing due
to oil leak:-

1.    Disconnect and remove damper
      from chassis. Bolt onto flat
      plate. do not hold directly
      in vice.

2.    Remove bolt from intermediate
      lever, the two nuts and washers
      from gland cover and withdraw
      main lever from casing complete
      with gland assembly.

To Replace:

1.    Place gland cover onto shaft
      of main lever followed by
      retaining ring, the new
      gland packing, which must be
      very carefully fitted over
      the shaft, pressure ring and
      gland spring.

      Gland packing - approximately
      2¼ ft. of 2 ply asbestos
      string prepared with Russian
      Tallow.

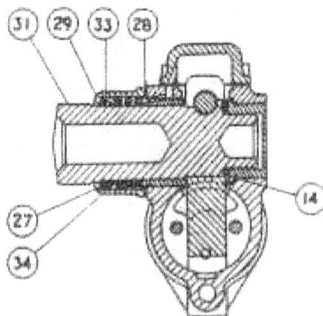

FIG. J22.   GLAND ASSEMBLY.

14.  Adjusting washer.
27.  Retaining ring.
28.  Joint washer.
29.  Gland cover.
33.  Gland packing.
34.  Pressure ring.

Place joint washer on casing, refit main lever, do not damage adjusting
washer. Press gland cover towards casing as far it will go and tighten
nuts with fingers only.

2.    The gland must only be tightened up when it has been warmed sufficiently
      to soften tallow impregnating packing. Heat end of shaft and the end of
      gland cover gently with blow lamp or gas flame. The temperature is
      approximately correct when it is just possible to bear the hand on end
      of shaft.

      Progressively tighten gland cover nuts. Excess tallow will be exuded
      and all space in gland filled.

3.    Refit bolt to intermediate lever. Fill damper with oil, expel air and
      refit to chassis.
      Refit bearing bolt to top of connecting link and main lever of damper.
      (The nut must only be tightened when the rear springs are taking full
      weight of body). This instruction also applies to nut of bearing bolt
      at lower end of connecting link and also nut securing Silentbloc
      bracket to rear spring plate.

4.    Reconnect oil pressure pipe to damper and expel air:-

      Start engine and run slowly in top gear approximately 10 miles per
      hour, move ride control lever to "Hard", remove air release plug on
      damper end cover, continue running until continous flow of oil from
      release plug.

      Replace plug and repeat for opposite side damper. Top up gearbox.

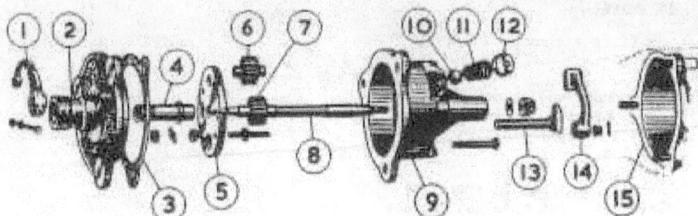

FIG. J23.  DAMPER CONTROL PUMP.

| | |
|---|---|
| 1. External operating lever. | 9. Housing. |
| 2. Cover. | 10. Valve ball. |
| 3. Vellumoid washer. | 11. Valve spring. |
| 4. Bush. | 12. Valve spring cover. |
| 5. Disc. | 13. Operating pin. |
| 6. Driven gear. | 14. Internal operating lever. |
| 7. Driving gear. | 15. Gearbox. |
| 8. Gear shaft. | |

### Testing "Ride Control":

Jack up rear wheel, engage top gear and run engine at approximate speed of 10 - 15 miles per hour.

Connect oil pressure gauge to plug on 4 way connection in the frame cruciform, just behind gearbox.

With hand lever in "Hard" position, adjust control rod so that oil pressure of 29 - 31 lbs/sq.in. is obtained.  Check with hand lever at "Normal" position that pressure is not less than 2½ lbs/sq.in.

SERVICE HANDBOOK

SILVER WRAITH — SILVER DAWN — BENTLEY MK. VI.
R. TYPE BENTLEY — PHANTOM IV.

# SECTION
# K
# HUBS, WHEELS AND TYRES

SECTION X.

HUBS, WHEELS AND TYRES

FRONT HUBS - WHEELS - TYRE PRESSURES.

SILVER WRAITH — SILVER DAWN — BENTLEY MK. VI.
R. TYPE BENTLEY — PHANTOM IV.

SECTION  K.

HUDS, WHEELS AND TYRES

SECTION K.

HUBS, WHEELS AND TYRES

FRONT HUBS:

Early series cars incorporated two single row journal ball bearings as shown in Fig. K1. As from Silver Wraith WME-1, Silver Dawn SCA-1 and Bentley B-1-GT, these ball bearings were replaced by two roller bearings as shown in Fig. K2. At the same time the method of wheel balancing was also changed, i.e., from a series of weighting washers in four fixed positions, to movable weights in variable positions.

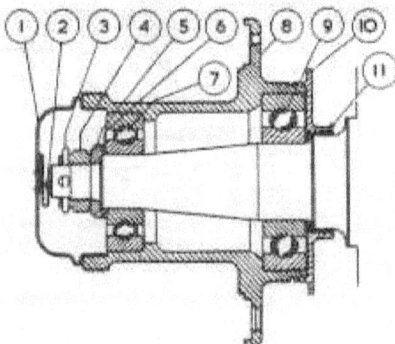

FIG. K1. HUB WITH BALL BEARING.

1. Dust cover.
2. Earth contact.
3. Split pin.
4. Castellated nut.
5. Key washer.
6. Adjusting washer.

7. Bearing (outer).
8. Hub.
9. Bearing (inner).
10. Grease retainer.
11. Distance piece.

To Remove Front Hub:

1. Remove wheel, slacken off brake adjusters and with two screwdrivers, remove dust cover.

2. Early type - remove sealing strip and split pin (3, Fig. K1), remove retaining nut (4), washer (5), and adjusting washer (6). Note nut on R.H. axle has R.H. thread and nut on L.H. axle has L.H. thread.

## SILVER WRAITH — SILVER DAWN — BENTLEY MK. VI.
## R. TYPE BENTLEY — PHANTOM IV.

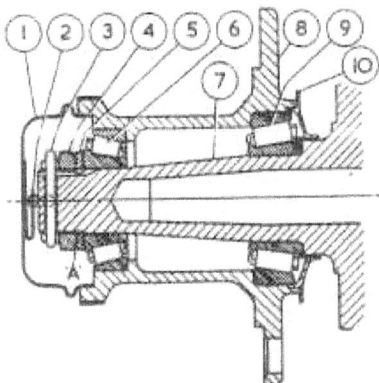

FIG. F2. HUB WITH ROLLER BEARINGS

1. Dust cover.
2. Earth contact.
3. Split pin.
4. Castellated nut.
5. Key washer.
6. Roller bearing (outer).
7. Shaft.
8. Hub.
9. Roller bearing (inner).
10. Grease retainer.

Screw extractor 3752/T1005 onto hub and withdraw hub complete with brake drum.

2a. Later type - remove sealing strip and split pin (3, Fig. K2) the nut (4) and washer (5), and withdraw hub by hand.

Normally the hub will withdraw complete with outer and inner taper roller bearing. Should the inner race of the inner bearing remain on stub axle, it will be found that grease retainer has parted from the hub. Prise inner race off stub axle with levers. Care must be taken not to damage roller cage or edges of grease retainer.

Note:- grease retainers are distinctly marked for assembly - "Off-side - Right-hand", or "Near-side - Left-hand".

3. To remove ball bearings from front hub, use two levers under lip of grease retaining cover (10, Fig. K1) and squarely prise off. Remove cover, this has an interference fit of .0023" to .0053". Covers have handed Acme threads and are clearly marked for re-assembly.

Pass a hardwood drift through inner bearing and tap out outer bearing, then remove inner bearing.

SILVER WRAITH — SILVER DAWN — BENTLEY MK. VI.

R. TYPE BENTLEY — PHANTOM IV.

To Refit Front Hub:

Only the correct amount of grease should be packed in each hub:-

Early type    -    1.6 ozs. Retinax A.
Later type    -    2.5 ozs. Retinax A.

Early Type:

1.    Refit hub and brake drum assembly, ensure this is driven fully
home, inner face of inner ball bearing contacts against distance
piece (11. Fig. K1).

Replace adjusting washer (6) and plain washer (5) and tighten
retaining nut.

Note:- adjusting washer (6) should just be rotatable by hand
       with nut fully tightened.

Later Type:

1.    Place hub in position, followed by outer bearing. Fit key washer
(5, Fig. F2) and tighten castellated nut sufficient to take up end
float in hub.
Unscrew nut sufficient to allow two .002" feelers to be inserted,
diametrically opposite one another, between nut and key washer "A",
Fig. K2. With feelers in position tighten nut hand tight with box
spanner, and split pin to lock, to give end float of .002" to .004".

WHEELS:

SILVER WRAITH    - either 17" x 5" or 16" x 6.50" well-base rims.
SILVER DAWN      - 16" x 5" - well-base rims.
BENTLEY          - 16" x 5" - well-base rims.
PHANTOM IV       - 17" x 5" - semi-drop centre rims.

TYRE PRESSURES:

The tables below show the correct method of determining tyre pressures,
according to axle weights, in pounds.

To determine the correct tyre pressure, weigh the car, each axle in
turn, with not more than 5 gallons of petrol in the tank, but otherwise in kerbside condition. Read off on the appropriate table, the tyre pressure against
the axle weight and size of tyre. If the tank is full, deduct 140 lbs. from the
rear axle weight of Bentley Mk.VI, Bentley "R" type, Silver Dawn and Silver Wraith
and 190 lbs. for Phantom IV, and add 20 lbs. to the front axle weights.

The tables do not apply to the Bentley Continental Sports Saloon, which
is fitted with Special High Speed Tyres.

Filling the petrol tank reduces the front axle load by about 20 lbs.
and passengers increase it by 150 lbs. The loaded front axle weight is assumed
to be 130 lbs. greater than normal kerbside weight.

Filling the petrol tank adds about 140 lbs. to the rear axle weight, 4 passengers will add another 450 lbs. making a total of 590 lbs. No allowance is made for a 5th passenger or for luggage. The loaded axle weight is assumed to be 600 lbs. above the normal kerbside condition, except in the case of Phantom IV when allowance is increased by 650 lbs. because of the larger petrol tank.

Obviously cars will usually run under this assumed weight, but some may often exceed it. It gives about the highest pressure which can be used for average conditions without undue harshness.

If a 5th passenger and 100-200 lbs. of luggage are carried, the rear tyre pressure should be increased by 3 lbs/sq.in. for better handling, and under Continental conditions, perhaps for better tyre life.

On limousines when 6 people are carried, the pressure should be increased by 4 lbs/sq.in.

SILVER WRAITH — SILVER DAWN — BENTLEY MK. VI.

R. TYPE BENTLEY — PHANTOM IV.

6.50" x 16" TYRES

| Pressure lbs/sq.in. | Pressure kg/sq.cm. | Weights - lbs. | |
|---|---|---|---|
| | | Front | Rear |
| 15 | 1.05 | | |
| 16 | 1.12 | | |
| 17 | 1.19 | | |
| 18 | 1.27 | | |
| 19 | 1.34 | | |
| 20 | 1.41 | 1722 - 1790 | |
| 21 | 1.48 | 1791 - 1860 | |
| 22 | 1.55 | 1861 - 1930 | |
| 23 | 1.62 | 1931 - 2000 | |
| 24 | 1.69 | 2001 - 2070 | |
| 25 | 1.76 | 2071 - 2140 | |
| 26 | 1.83 | 2141 - 2210 | |
| 27 | 1.90 | | |
| 28 | 1.97 | | 1745 - 1810 |
| 29 | 2.04 | | 1811 - 1880 |
| 30 | 2.11 | | 1881 - 1950 |
| 31 | 2.18 | | 1951 - 2020 |
| 32 | 2.25 | | 2021 - 2090 |
| 33 | 2.32 | | 2091 - 2160 |
| 34 | 2.39 | | 2161 - 2230 |
| 35 | 2.46 | | 2231 - 2300 |
| 36 | 2.53 | | 2301 - 2370 |
| 37 | 2.60 | | |
| 38 | 2.67 | | |
| 39 | 2.74 | | |
| 40 | 2.81 | | |

SILVER WRAITH — SILVER DAWN — BENTLEY MK. VI.

R. TYPE BENTLEY — PHANTOM IV.

7.00" x 16" TYRES

| Pressure lbs/sq.in. | Pressure Kg/sq.cm. | Weights - lbs. | |
|---|---|---|---|
| | | Front | Rear |
| 15 | 1.05 | | |
| 16 | 1.12 | | |
| 17 | 1.19 | | |
| 18 | 1.27 | | |
| 19 | 1.34 | | |
| 20 | 1.41 | 1921 - 1990 | |
| 21 | 1.48 | 1991 - 2060 | |
| 22 | 1.55 | 2061 - 2130 | |
| 23 | 1.62 | 2131 - 2200 | |
| 24 | 1.69 | 2201 - 2270 | |
| 25 | 1.76 | 2271 - 2340 | |
| 26 | 1.83 | 2341 - 2410 | |
| 27 | 1.90 | 2411 - 2480 | |
| 28 | 1.97 | | 2011 - 2080 |
| 29 | 2.04 | | 2081 - 2150 |
| 30 | 2.11 | | 2151 - 2220 |
| 31 | 2.18 | | 2221 - 2290 |
| 32 | 2.25 | | 2291 - 2360 |
| 33 | 2.32 | | 2361 - 2430 |
| 34 | 2.39 | | 2431 - 2500 |
| 35 | 2.46 | | 2501 - 2570 |
| 36 | 2.53 | | 2571 - 2640 |
| 37 | 2.60 | | |
| 38 | 2.67 | | |
| 39 | 2.74 | | |
| 40 | 2.81 | | |

**SILVER WRAITH — SILVER DAWN — BENTLEY MK. VI.**
**R. TYPE BENTLEY — PHANTOM IV.**

### 7.50" x 16" TYRES

| Pressure lbs/sq.in. | Pressure Kg/sq.cm. | Weights - lbs. Front | Rear |
|---|---|---|---|
| 15 | 1.05 | 1921 - 1995 | |
| 16 | 1.12 | 1996 - 2070 | |
| 17 | 1.19 | 2071 - 2145 | |
| 18 | 1.27 | 2146 - 2220 | |
| 19 | 1.34 | 2221 - 2295 | |
| 20 | 1.41 | 2296 - 2370 | |
| 21 | 1.48 | 2371 - 2445 | 1901 - 1975 |
| 22 | 1.55 | | 1976 - 2050 |
| 23 | 1.62 | | 2051 - 2125 |
| 24 | 1.69 | | 2126 - 2200 |
| 25 | 1.76 | | 2201 - 2275 |
| 26 | 1.83 | | 2276 - 2350 |
| 27 | 1.90 | | 2351 - 2425 |
| 28 | 1.97 | | 2426 - 2500 |
| 29 | 2.04 | | 2501 - 2575 |
| 30 | 2.11 | | 2576 - 2650 |
| 31 | 2.18 | | 2651 - 2725 |
| 32 | 2.25 | | 2726 - 2800 |
| 33 | 2.32 | | 2801 - 2875 |
| 34 | 2.39 | | 2876 - 2950 |
| 35 | 2.46 | | 2951 - 3025 |
| 36 | 2.53 | | 3026 - 3100 |
| 37 | 2.60 | | 3101 - 3175 |
| 38 | 2.67 | | 3176 - 3250 |
| 39 | 2.74 | | |
| 40 | 2.81 | | |

SILVER WRAITH — SILVER DAWN — BENTLEY MK. VI.

R. TYPE BENTLEY — PHANTOM IV.

6.50" x 17" TYRES

| Pressure lbs/sq.in. | Pressure kg/sq.cm. | Weights - lbs. | |
|---|---|---|---|
| | | Front | Rear |
| 15 | 1.05 | | |
| 16 | 1.12 | | |
| 17 | 1.19 | | |
| 18 | 1.27 | | |
| 19 | 1.34 | | |
| 20 | 1.41 | 1810 - 1884 | |
| 21 | 1.48 | 1885 - 1959 | |
| 22 | 1.55 | 1960 - 2034 | |
| 23 | 1.62 | 2035 - 2109 | |
| 24 | 1.69 | 2110 - 2184 | |
| 25 | 1.76 | 2185 - 2259 | |
| 26 | 1.83 | 2260 - 2334 | |
| 27 | 1.90 | 2335 - 2409 | 1865 - 1939 |
| 28 | 1.97 | | 1940 - 2014 |
| 29 | 2.04 | | 2015 - 2089 |
| 30 | 2.11 | | 2090 - 2164 |
| 31 | 2.18 | | 2165 - 2239 |
| 32 | 2.25 | | 2240 - 2314 |
| 33 | 2.32 | | 2315 - 2389 |
| 34 | 2.39 | | 2390 - 2464 |
| 35 | 2.46 | | 2465 - 2539 |
| 36 | 2.53 | | 2540 - 2614 |
| 37 | 2.60 | | 2615 - 2689 |
| 38 | 2.67 | | 2690 - 2764 |
| 39 | 2.74 | | |
| 40 | 2.81 | | |

7.00" x 17" TYRES

| Pressure lbs/sq.in. | Pressure kg/sq.cm. | Weights - lbs. Front | Rear |
|---|---|---|---|
| 15 | 1.05 | 1843 - 1936 | |
| 16 | 1.12 | | |
| 17 | 1.19 | | |
| 18 | 1.27 | | |
| 19 | 1.34 | | |
| 20 | 1.41 | 1843 - 1936 | |
| 21 | 1.48 | 1937 - 2030 | |
| 22 | 1.55 | 2031 - 2124 | |
| 23 | 1.62 | 2125 - 2218 | |
| 24 | 1.69 | 2219 - 2312 | |
| 25 | 1.76 | 2313 - 2406 | |
| 26 | 1.83 | 2407 - 2500 | |
| 27 | 1.90 | 2501 - 2594 | 1901 - 1994 |
| 28 | 1.97 | | 1995 - 2088 |
| 29 | 2.04 | | 2089 - 2182 |
| 30 | 2.11 | | 2183 - 2276 |
| 31 | 2.18 | | 2277 - 2370 |
| 32 | 2.25 | | 2371 - 2464 |
| 33 | 2.32 | | 2465 - 2558 |
| 34 | 2.39 | | 2559 - 2652 |
| 35 | 2.46 | | 2653 - 2746 |
| 36 | 2.53 | | 2747 - 2840 |
| 37 | 2.60 | | 2841 - 2934 |
| 38 | 2.67 | | 2935 - 3028 |
| 39 | 2.74 | | 3029 - 3122 |
| 40 | 2.81 | | 3123 - 3216 |

**SILVER WRAITH — SILVER DAWN — BENTLEY MK. VI.**

**R. TYPE BENTLEY — PHANTOM IV.**

8.00" x 17" TYRES

| Pressure lbs/sq.in. | Pressure kg/sq.cm. | Weights - lbs. Front | Rear |
|---|---|---|---|
| 15 | 1.05 | | |
| 16 | 1.12 | | |
| 17 | 1.19 | | |
| 18 | 1.27 | | |
| 19 | 1.34 | | |
| 20 | 1.41 | | |
| 21 | 1.48 | | |
| 22 | 1.55 | | |
| 23 | 1.62 | | |
| 24 | 1.69 | | |
| 25 | 1.76 | 2378 - 2486 | |
| 26 | 1.83 | 2487 - 2594 | |
| 27 | 1.90 | 2595 - 2702 | |
| 28 | 1.97 | 2703 - 2810 | |
| 29 | 2.04 | 2811 - 2918 | |
| 30 | 2.11 | 2919 - 3026 | 2399 - 2506 |
| 31 | 2.18 | | 2507 - 2614 |
| 32 | 2.25 | | 2615 - 2722 |
| 33 | 2.32 | | 2723 - 2830 |
| 34 | 2.39 | | 2831 - 2938 |
| 35 | 2.46 | | 2939 - 3046 |
| 36 | 2.53 | | 3047 - 3154 |
| 37 | 2.60 | | 3155 - 3262 |
| 38 | 2.67 | | 3263 - 3370 |
| 39 | 2.74 | | 3371 - 3478 |
| 40 | 2.81 | | 3479 - 3586 |

SERVICE HANDBOOK

SILVER WRAITH — SILVER DAWN — BENTLEY MK. VI.
R. TYPE BENTLEY — PHANTOM IV.

# SECTION
# L
# CHASSIS LUBRICATION

## Service Handbook

SILVER WRAITH — SILVER DAWN — BENTLEY MK. VI.
R. TYPE BENTLEY — PHANTOM IV.

## SECTION L.

## CENTRALISED CHASSIS LUBRICATION SYSTEM

PUMP UNIT  -  PIPE LINE  -  DRIP PLUGS

# SERVICE HANDBOOK

### SILVER WRAITH — SILVER DAWN — BENTLEY MK. VI.
### R. TYPE BENTLEY — PHANTOM IV.

## SECTION L.

## CENTRALISED CHASSIS LUBRICATION SYSTEM

### List of Illustrations:

### SILVER WRAITH — SILVER DAWN — BENTLEY MK. VI.
### R. TYPE BENTLEY — PHANTOM IV.

## THE CENTRALISED CHASSIS LUBRICATION SYSTEM

GENERAL:

     A Luvax Bijur foot-operated pump and combined oil reservoir is located on the front of the dashboard and supplies oil under pressure for chassis lubrication.

     Diagrams of the system are given in Figures L1 and L2, the piping being coloured in red. Red discs indicate the positions of the drip plugs, and the rating of each is given in parenthesis against description of part lubricated.

THE PUMP UNIT:

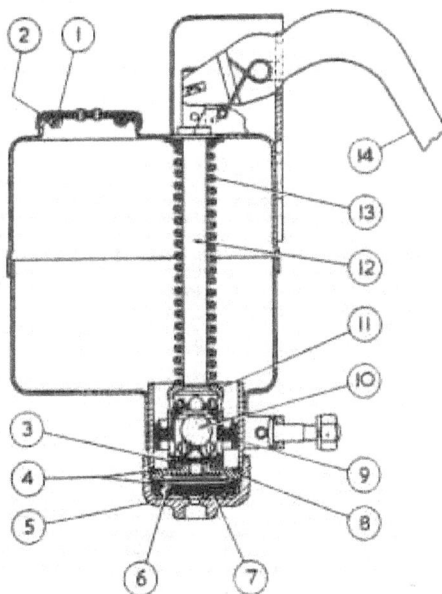

FIG. L3. CHASSIS LUBRICATION PUMP.

| | |
|---|---|
| 1. Filler cap. | 8. Strainer plate. |
| 2. Filler cap joint washer. | 9. Piston cup. |
| 3. Piston valve disc. | 10. Piston valve ball. |
| 4. Strainer plate joint washers. | 11. Piston rod valve unit. |
| 5. Cylinder cap nut. | 12. Piston rod. |
| 6. Strainer pad. | 13. Spring. |
| 7. Strainer support. | 14. Pedal. |

The construction of the pump unit is shown in Fig. L3.

The pedal is mounted on a fulcrum pin, and when depressed raises the piston in the cylinder and at the same time compresses the return spring.

As the piston rises, oil is drawn through a ball valve, in the centre of the piston, to the underside.

The pressure being removed from the pedal, the return spring forces the piston downwards. The pressure on the oil below the piston closes the ball valve, thus preventing a return flow through the piston, the oil being forced downwards through the filter and out of the pump outlet.

The spring is so rated that the pressure is practically constant throughout the stroke, and the time period during which the piston descends depends on the viscosity of the oil.

When the piston reaches the end of its stroke, it effectively closes the hole in the filter-retaining plate so that the oil cannot leak away while the pump is not being operated.

## PUMP FILTER:

If, with the pump unit correctly coupled up to the pipe lines, the pump lever fails to return to its normal position after being pressed down, it is probable that the filter has become clogged. Disconnect the chassis oil line at the pump outlets and unscrew the cylinder cap on the underside of the reservoir. Note the position in the cylinder cap of the filter retaining plate, with its gaskets, before dismantling. Lift out to expose the felt filter disc, discard this disc and replace with a new one. Beneath the felt disc in the cap is a wire gauze disc, this should be left in position with the ridges against the cap. When re-assembling, replace the filter retaining plate and gaskets, with the hollow side of the plate facing the felt disc. Be sure that both the gaskets are in the correct position.

After reconnecting the system, it should be primed until oil is exuding from each bearing.

## TESTING THE PUMP UNIT:

Disconnect the chassis oil line from the pump outlets, and close the outlets by screwing in solid plugs. Depress the pedal, with the outlets plugged, the piston should hold the pressure; if the pedal shows a visible upwards motion during a period of 2 minutes, it indicates a leak past the piston, either past the cup leathers or through the ball valve. Too thin an oil in the reservoir will also give this effect, therefore examine the oil and if necessary replace with the recommended grade.

## THE OIL PIPE LINE:

Brass tubing of 5/32" outside diameter is used for the oil pipe line, and all connections are made with screwed joints. The connections from the pipe line to the junction-pieces and drip plugs are made with compression sleeves (or olives), which are permanently pinched onto the end of the tube as the nut is tightened up. Flexible connections are used between the frame and the axles.

## THE DRIP PLUGS:

In each drip plug is an accurate restriction orifice which controls the flow of oil to the bearing, and also a valve which prevents oil draining away from the system when the car is at rest. The plugs are stamped with a letter and number indicating the shapes and relative rates of oil emission respectively, a higher number indicating a greater rate.

The drip plugs never require cleaning, and being non-adjustable and non-demountable, no attempt must be made to take them apart. If one is suspected of being defective, it must be replaced with a new one of the same rating.

SILVER WRAITH — SILVER DAWN — BENTLEY MK. VI.
R. TYPE BENTLEY — PHANTOM IV.

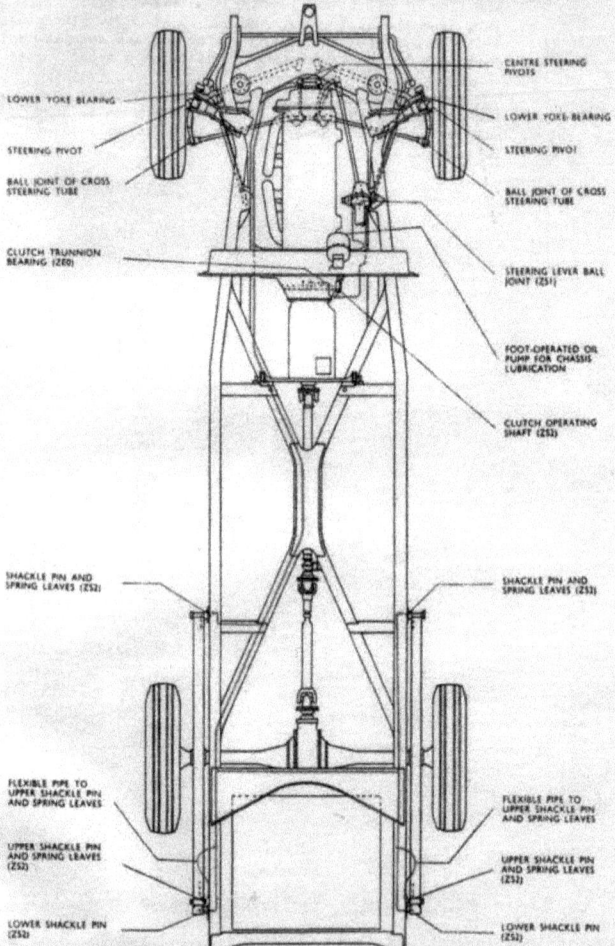

CENTRE STEERING PIVOTS

LOWER YOKE BEARING

LOWER YOKE BEARING

STEERING PIVOT

STEERING PIVOT

BALL JOINT OF CROSS STEERING TUBE

BALL JOINT OF CROSS STEERING TUBE

CLUTCH TRUNNION BEARING (Z60)

STEERING LEVER BALL JOINT (Z51)

FOOT-OPERATED OIL PUMP FOR CHASSIS LUBRICATION

CLUTCH OPERATING SHAFT (Z52)

SHACKLE PIN AND SPRING LEAVES (Z52)

SHACKLE PIN AND SPRING LEAVES (Z52)

FLEXIBLE PIPE TO UPPER SHACKLE PIN AND SPRING LEAVES

FLEXIBLE PIPE TO UPPER SHACKLE PIN AND SPRING LEAVES

UPPER SHACKLE PIN AND SPRING LEAVES (Z52)

UPPER SHACKLE PIN AND SPRING LEAVES (Z52)

LOWER SHACKLE PIN (Z52)

LOWER SHACKLE PIN (Z52)

Fig. 1.—DIAGRAM OF CHASSIS LUBRICATION SYSTEM.
MANUAL GEARBOX

SILVER WRAITH — SILVER DAWN — BENTLEY MK. VI.

R. TYPE BENTLEY — PHANTOM IV.

CENTRE STEERING
LEVER PIVOT AND
BALL JOINTS

LOWER YOKE BEARING

STEERING PIVOT

BALL JOINT OF CROSS
STEERING TUBE

LOWER YOKE BEARING

STEERING PIVOT

BALL JOINT OF CROSS
STEERING TUBE

STEERING LEVER BALL
JOINT (ZS1)

SHACKLE PIN AND
SPRING LEAVES (ZS2)

SHACKLE PIN AND
SPRING LEAVES (ZS2)

FLEXIBLE PIPE TO
UPPER SHACKLE PIN
AND SPRING LEAVES

UPPER SHACKLE PIN
AND SPRING LEAVES
(ZS2)

LOWER SHACKLE PIN
(ZS2)

FLEXIBLE PIPE TO
UPPER SHACKLE PIN
AND SPRING LEAVES

UPPER SHACKLE PIN
AND SPRING LEAVES
(ZS2)

LOWER SHACKLE PIN
(ZS2)

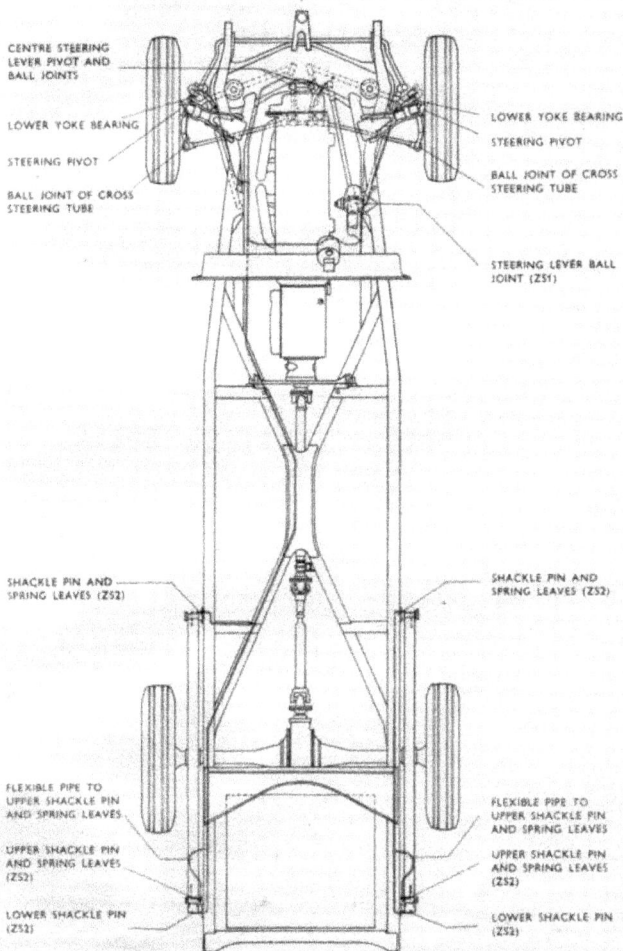

Fig. 2—DIAGRAM OF CHASSIS LUBRICATION SYSTEM.

AUTO GEARBOX

SERVICE HANDBOOK

SILVER WRAITH — SILVER DAWN — BENTLEY MK. VI.
R. TYPE BENTLEY — PHANTOM IV.

# SECTION
# M
# ELECTRICAL

SILVER WRAITH — SILVER DAWN — BENTLEY MK. VI.

R. TYPE BENTLEY — PHANTOM IV.

SECTION M.

E L E C T R I C A L .

List of Illustrations:-

# SERVICE HANDBOOK

SILVER WRAITH — SILVER DAWN — BENTLEY MK. VI.
R. TYPE BENTLEY — PHANTOM IV.

SECTION M.

E L E C T R I C A L

BATTERY - DYNAMO - VOLTAGE REGULATOR AND CUT-OUT -
FUSES - STARTER MOTOR AND DRIVE - STARTER MOTOR SWITCH
- MICRO-SWITCH - IGNITION DISTRIBUTOR - IGNITION COIL
- SPARKING PLUGS - HEADLAMPS - PASS LAMPS - SIDE
LAMPS - STOP AND TAIL LAMP - NUMBER PLATE AND REVERSE
LIGHTS - SWITCHBOX - HORNS - TRAFFICATORS - DIRECTION
INDICATOR LIGHTS - WINDSCREEN WIPERS.

SILVER WRAITH — SILVER DAWN — BENTLEY MK. VI.

R. TYPE BENTLEY — PHANTOM IV.

SECTION M

ELECTRICAL

The electrical system is earthed on the positive side of the battery to the chassis frame, and all switching is done in the negative leads.

The various wiring diagram, show the units with their electrical connections, the wiring being indicated in colours to correspond with actual coverings.

BATTERY:

The following are the recommended battery types:-

| Battery Maker's Designation | | Voltage | Normal Charging Current |
|---|---|---|---|
| P & R Dagenite | Exide | | |
| 6 HZP9-S | 6 MXP9-R | 12 | 5 amperes |

The specific gravity figures given below apply to both makes of batteries:-

| Climate | Specific Gravity of Sulphuric Acid Solution (Corrected to 70°F) | |
|---|---|---|
| | Filling in for First Charge 6 HZP9-S 6 MXP9-R | Fully Charged |
| Temperate | 1.260 | 1.280 (1.270 - 1.285) |
| Tropical (i.e. where temperature is frequently over 90°F. | 1.190 | 1.210 (1.200 - 1.215) |

Battery terminals should be kept clean and well coated with Lanolin or Vaseline. Remove corrosion with a solution of ammonium carbonate, applying with a rag.

DYNAMO:

On early models, a 4½" Lucas dynamo was fitted, then a 5" Lucas dynamo; both of these were of the non-insulated type, i.e., the dynamo frame constituted the earth connection. Later, this was superseded by the Lucas 5" fully-insulated type, i.e., the dynamo earth connection being taken to a separate insulated terminal on the dynamo frame, which is suitably connected to chassis and this is now current production.

### SILVER WRAITH — SILVER DAWN — BENTLEY MK. VI.
### R. TYPE BENTLEY — PHANTOM IV.

<u>Performance Data</u>:   - RA.5 Dynamo.

| | |
|---|---|
| Cutting-in speed | - 800 - 850 r.p.m. at 13.0 dynamo volts. |
| Output | - 24 amps. at 1400 - 1550 r.p.m. at 13.5 dynamo volts. (Reading taken with regulator disconnected and using .55 ohm resistance load.) |
| Rotation | - Clockwise, from driving end. |
| Brush Spring Tension | - 15 - 25 ozs. |
| Armature Resistance | - 0.19 ohms. |
| Field Coil Resistance | - 6 ohms total. |
| Suppression | - 1 - mfd condenser. |

<u>Dynamo Belt Tension</u>:

The dynamo is driven by a Vee belt, off the crankshaft pulley, this belt also driving the water pump and fan.

The tension of the belt should be such that it can be moved transversely with the fingers through a total distance of 1" (i.e., ½" in each direction) when checked at a point equidistant between the crankshaft and fan pulleys, see Fig. M1.

To adjust belt, slacken dynamo securing nuts A, B, C & D and move dynamo towards or away from cylinder block as required.

To remove belt, slacken dynamo and move hard-up to cylinder block. Do not strain belts over pulleys and do not forcibly turn fan blades by hand.

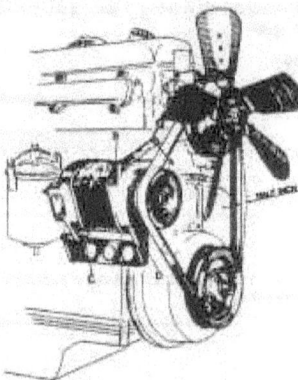

FIG. M1.   DYNAMO BELT ADJUSTMENT.

<u>Testing Dynamo in Position</u>:

1.   Inspect driving belt and adjust if necessary.

2.   Check that dynamo and control box are correctly connected, i.e., dynamo terminals "F" and "D" respectively. If terminals are not marked, the "F" or field terminal can be identified as the small terminal.

3.   With all lights and accessories switched off, disconnect wires from dynamo "F" and "D" terminals, connect terminals together with a short piece of wire, start engine and run at normal idling speed. Connect a 20 volt full scale moving coil voltmeter with its negative lead to terminal "D" and the positive lead to the chassis frame.

Gradually increase engine speed, when voltmeter reading should rise rapidly without fluctuation. No attempt must be made to force up reading by racing engine, 800 r.p.m. is sufficient.

4.   A low reading (approximately, 1 volt) may indicate a fault in the field winding. A reading of 5 volts indicates a fault either in the armature or brushes.

5.  Remove brush gear cover plate, hold back each brush spring, and move brush by gently pulling on its flexible connection. If movement is sluggish, remove brush and clean.

    Brushes must always be replaced in original positions. If the brushes are so worn that they will no longer bed correctly on the commutator, or if core of flexible connector is exposed on brush face, new brushes must be fitted. If commutator is dirty, clean with petrol moistened cloth.

6.  If a fault in the suppression condenser is suspected, this is most unlikely on the later fully insulated type, disconnect condenser and retest, if voltage now builds up, condenser is at fault. If there is no reading or a very low one, dynamo must be dismantled.

Dismantling:

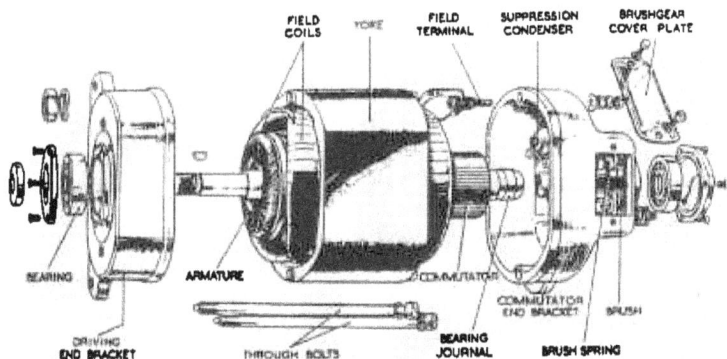

FIELD COILS  YOKE  FIELD TERMINAL  SUPPRESSION CONDENSER  BRUSHGEAR COVER PLATE

BEARING  ARMATURE  COMMUTATOR  COMMUTATOR END BRACKET  BRUSH

DRIVING END BRACKET  THROUGH BOLTS  BEARING JOURNAL  BRUSH SPRING

FIG. M2.  DYNAMO - DISMANTLED.

Remove dynamo from engine, and before dismantling, mark yoke and end brackets to ensure correct alignment on assembly.

1.  Take off drive pulley. Remove key from armature shaft.

2.  Remove brush gear cover plates, hold back brush spring and slide brushes from holders. Disconnect field terminal. Unscrew and withdraw the two through bolts securing end brackets to yoke.

3.  Pull commutator end bracket off yoke, bracket locates on two loose dowels passing through laminated yoke and is fitted with roller bearing for armature shaft. The driving end bracket and armature can now be lifted out. The end bracket carries the commutator shaft in a ball bearing which should not be needlessly disturbed.

    If the bearing or armature requires replacement, press out shaft from end bracket using hand press.

Re-assembly:

      Re-assembling dynamo is approximately reverse of dismantling.

1. The end brackets locate over the dowel pins in the yoke, and must be replaced in original positions, relative to the yoke and each other.

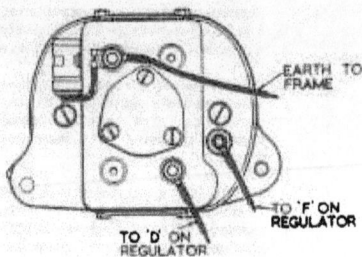

FIG. M3. DYNAMO CONNECTIONS.

2. Ensure that square insulating washers on field terminal fit correctly.

3. The distance piece on collar at the driving end must be replaced with chamfered edge towards machine.

Dynamo Earthing:

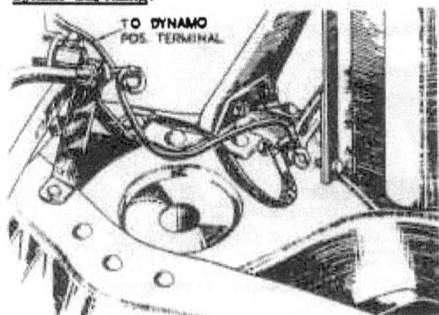

FIG. M4. DYNAMO EARTH CONNECTION.

FIG. M5. RADIATOR BONDING STRIPS.

      It is essential that the dynamo is correctly earthed to obviate any electrolytic action in the cooling system which may carry an excessive deposit of oxides to the radiator.

      In cases where this trouble has been experienced, Figs. M4 & M5, show the service scheme for correct earthing of dynamo and bonding of matrix. On the early type of non-insulated dynamo, Tufnol bushes are now fitted to insulate the dynamo from the carrier bracket, and a separate earthing strip to the chassis frame is fitted.

VOLTAGE REGULATOR AND CUT-OUT:

      The control unit also incorporates a choke condenser filter to prevent radio interference.

To Check and Adjust Regulator:

1. Withdraw cables from terminal "A1", and "A" if connected.

SILVER WRAITH — SILVER DAWN — BENTLEY MK. VI.

R. TYPE BENTLEY — PHANTOM IV.

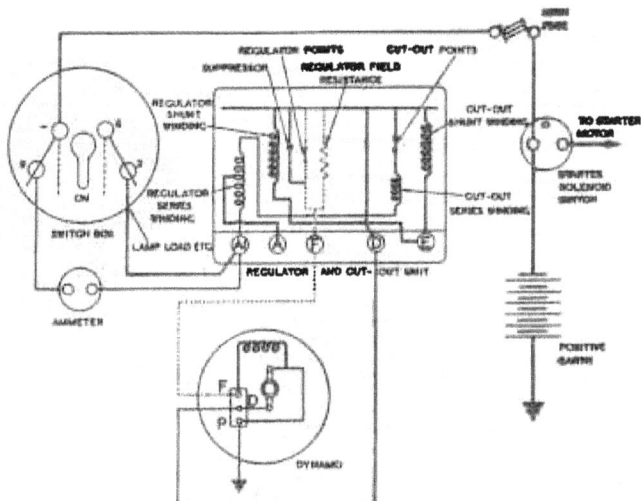

FIG. M6.  CHARGING CIRCUIT.

2.  Connect negative lead of a moving coil (0 - 20 volt) voltmeter to terminal "D", and the positive to earth.

3.  Start engine and increase speed until voltmeter needle flicks and then steadies.  This should occur between limits below.

| Air Temperature - at Regulator | Setting |
|---|---|
| 50°F | 16.0 - 16.6 volts. |
| 68°F | 15.7 - 16.3 volts. |
| 86°F | 15.4 - 16.0 volts. |
| 104°F | 15.1 - 15.9 volts. |

If the voltage at which the reading steadies is outside these limits, adjust regulator as under.

4.  Hold adjusting screw (B, Fig. M7) and slacken off locknut "A".  Turn adjusting screw clockwise to raise the setting, or anti-clockwise to lower.

Only a very small movement is necessary.
Re-tighten locknut and re-test.

When adjusting, do not run engine at more than half throttle or a false reading will be shown.

### SILVER WRAITH — SILVER DAWN — BENTLEY MK. VI.
### R. TYPE BENTLEY — PHANTOM IV.

Cleaning Regulator Contacts:

    To clean the vibrating contacts, unscrew "C" and "D", Fig. M7, and polish with fine emery cloth.

Cleaning and Setting Out-out Contacts:

    Check out-out contacts are clean and making good contact when closed.

    To clean, insert strip of fine glass paper between contacts, holding contacts closed by hand, draw paper through several times.

    Check voltage at which out-out contacts close by connecting voltmeter between terminals "D" and "E", and raising engine speed. When voltage reaches 12.7 to 13.3, contacts should close.

    Adjust by means of screw "F".

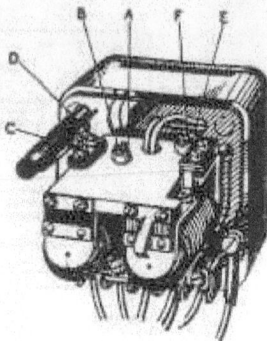

FIG. M7.   REGULATOR CONTACTS.

FUSES:

    The circuit fuses are one strand of No. 32 S.W.G. tinned copper wire.

    The main fuse is 3 strands of No. 32 tinned copper wire. On early models, the main fuse is contained in the circuit fuse box, and on later models, the main fuse is in a separate box at the side of the circuit box.

STARTER MOTOR AND DRIVE:

    The Starter motor and drive consists of a Lucas starter motor with a Rolls-Royce reduction gear and drive unit.

    The reduction gear provides a gear ratio between motor and crankshaft of 16 : 1.

Testing Starter Motor in Position:

    Switch on lamps and operate starter. If lights dim, but starter does not operate, suspect starter pinion jammed with flywheel gear ring or broken brush connection. In either case, the starter must be removed for examination.

    Should the lamp retain full brilliance when starter is switch is operated, check switch, and if in order, examine all connections.

    Starter drive clutch slip is indicated if the starter motor operates, but does not crank engine.

To Remove the Starter Motor:

1.    Disconnect positive lead from battery and lead from starter motor.

2.    Remove the four long setscrews securing motor to the clutch casing.

SILVER WRAITH — SILVER DAWN — BENTLEY MK. VI.

R. TYPE BENTLEY — PHANTOM IV.

3.    Withdraw the drive unit from the rear of the clutch casing and the motor from the front.

To refit, reverse the above procedure.

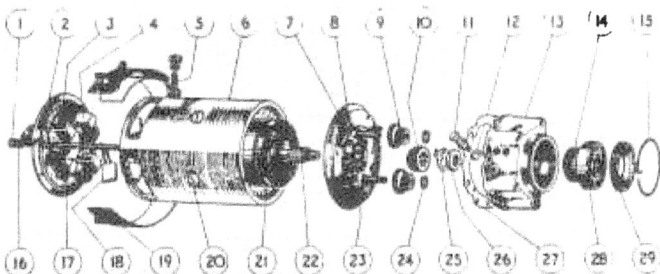

FIG. M8.   STARTER MOTOR AND REDUCTION GEAR.

1.  Through bolt.
2.  Commutator end bracket.
3.  Brush spring.
4.  Brush holder.
5.  Terminal.
6.  Yoke.
7.  Driving end bracket.
8.  Annular support - Gears.
9.  Compound pinion.
10. Driving gear.
11. Plug, lubricator.
12. Joint washer.
13. Gear housing.
14. Driven gear.
15. Locking ring.
16. Flat spring washer.
17. Ball or roller bearing.
18. Brush.
19. Cover band and cork joint.
20. Retaining screw - field coil.
21. Field coil.
22. Armature shaft.
23. Ball bearing.
24. Adjusting washer - compound pinion.
25. Lock washer - armature shaft.
26. Nut - armature shaft.
27. Joint washer (Vellumoid).
28. Ball bearing - gear housing.
29. Retaining nut - gear housing.

Examination of Commutator or Brushes:

Remove the cover band (19, Fig. M8), hold back each of the brush springs (3) and move the brush. If sluggish, remove brush (4) and ease sides with smooth file.

Test the brush spring tension with a spring balance, the correct tension is 30 - 40 ozs.

If new brushes are fitted, these are pre-formed so that no hand bedding is necessary. It is advisable to run these in by running motor for 15 minutes at 6 volts.

Clean commutator with petrol moistened cloth.

Secure the starter motor in a vice and test with 12 volt battery. If there is evidence of an internal fault, replace with a service unit.

SILVER WRAITH — SILVER DAWN — BENTLEY MK. VI.

R. TYPE BENTLEY — PHANTOM IV.

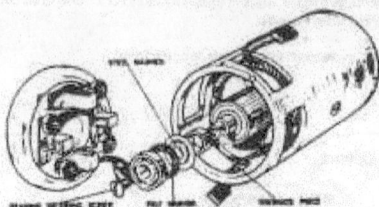

FIG. M9. COMMUTATOR END BEARING.

Starter Drive Gearbox:

Remove the aluminium casing, (13, Fig. M8) and the two compound pinions(9)

Examine the gears; if lubrication has been neglected, they may have picked up. Examine bronze bushes for scoring or seizure on pins. Check oil holes.

Pack bearings with H.M.P. grease and when re-assembling, note co-relation marks etched on teeth to ensure correct meshing in original positions.

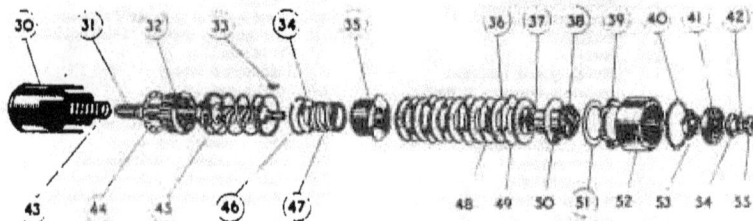

FIG. M10. DRIVE DISMANTLED.

| | | | |
|---|---|---|---|
| 30. Shell. | | 43. Disengaging spring. | |
| 31. Shaft. | | 44. Fibre washer. | |
| 32. Pinion. | | 45. Engaging spring. | |
| 33. Woodruff Key. | | 46. Spring ring. | |
| 34. Washer. | | 47. Damping spring. | |
| 35. End clutch plate and damping spring housing. | | 48. Clutch disc with spigot. | |
| | | 49. Clutch disc with projections. | |
| 36. Clutch disc - Ferodo. | | 50. Outer clutch disc. | |
| 37. Spring ring. | | 51. Fibre distance washer. | |
| 38. Operating nut. | | 52. Cover. | |
| 39. Clutch ring. | | 53. Stop operating bush. | |
| 40. Locking ring. | | 54. Plain washer. | |
| 41. Ball bearing. | | 55. Nut. | |
| 42. Lock washer. | | | |

- M.8 -

Starter Drive:

       Dismantling:-

1.    Remove small aluminium cover.

2.    Mount drive vertically in vice, remove lock washer (42, Fig. M10) and ring nut (55).

       Support the housing and drive out the assembly with an aluminium drift.

3.    Remove the stop (53) and the locking ring (40).

Re-assembly:

1.    Place fibre washer (44) over pinion (32), chamfer outermost. Place pinion in shell (30) and drop shaft (31) into shell through pinion, followed by engaging spring (45).

2.    Place the assembly, consisting of end clutch plate (35), damping spring (47), washer (34), and spring ring (46), over shaft and into engaging spring and shell.

3.    Assemble clutch discs, hold together with fingers, remove operating nut (38) and measure overall thickness of discs, this should be 1.108" - .010".

       The normal thickness of each Ferodo disc is .094". If above limit, rub down, using medium glass paper. If below limit increase by selective assembly of Ferodo discs.

4.    Soak the discs in engine oil for 30 minutes, then place the discs and operating nut over shaft, force down against spring by screwing the shaft up through the operating nut, and continue assembly.

5.    Place fibre distance washer (51) into the clutch ring (39), chamfered outer diameter to ring, then place the ring and distance washer assembly on to the outer clutch disc (50) and hold down by turning shaft. Next fit cover (52) and retain with locking ring (40).

6.    With stop operating bush (53) in position, key (33) in keyway, replace aluminium housing and bearing. Secure with plain washer (54) locking washer (42) and slotted ring nut (55). Lubricate bearing with H.M.P. grease and replace end cover.

Checking Clutch Slip:

       The clutch must be set to slip within the limits of 15 to 35 lbs/ft.

       To check with torque spanner, remove slotted ring nut (55) from the end of the drive and substitute ½" B.S.F. nut, to allow for standard adaptor.

       Mount drive vertically in vice and check slip with torque spanner. If torque spanner is not available, see below.

### SILVER WRAITH — SILVER DAWN — BENTLEY MK. VI.
### R. TYPE BENTLEY — PHANTOM IV.

Locally manufacture a torque arm from
a piece of mild steel 15" long (approximately),
drilled and filed at one end to leave two or
three projections in the form of teeth to engage
with the teeth of the pinion (Fig. M10). A
small hole should be drilled at the other end,
to engage hook of spring balance, having 12"
between hole centres.

Check as shown in Fig. M11, using
spring balance at right angles to torque arm.

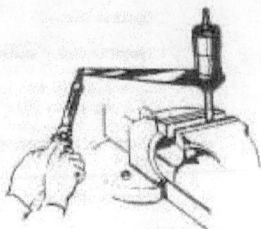

FIG. M11. CHECKING CLUTCH SLIP.

STARTER MOTOR SWITCH:

To test switch in position,
disconnect cables from switch, correct test
leads from 12 volt battery. The negative lead should be connected to terminal with
grub screw, the positive lead held against the switch body or fixing bracket.

If switch is working, it will be heard to operate every time circuit is
completed. If not, fit a replacement.

For emergency use or testing, the Solenoid may be manually operated by
pressing in the rubber cap, which covers an extension of the switch movement.

MICRO-SWITCH:

On cars fitted with the Automatic Gearbox, a small micro-switch, fitted
at the base of the steering column, is inserted in the starter motor circuit.
This switch is operated by the gear range selector lever, to ensure that the
engine can only be started with the gearbox in Neutral.

In cases of failure of the starter motor to operate, inspection should
be made that the gear range lever is definitely operating the toggle lever on the
switch. If necessary, move switch to reposition correctly on steering column.
Ensure that correct operation of reversing lights is not disturbed.

IGNITION DISTRIBUTOR:

The distributor is of the three lobe (four lobe on Phantom IV) cam and
twin contact breaker arm type. The automatic centrifugal advance mechanism is
housed in the base of the distributor.

The firing order (embossed on the cover) is 1, 4, 2, 6, 3, 5, for Silver
Wraith, Silver Dawn and Bentley, and 1, 6, 2, 5, 8, 3, 7, 4, for the Phantom IV.

The direction of rotation is clockwise, when viewed from the top.

Cleaning Contacts:

Lift rotor arm off spindle. Remove screws securing springs to anchorage.
Remove contact plate locking screws "B" & "E", Fig. M12., remove the contact plates
complete with breaker arm. The screws, G, H, & J, MUST NOT be disturbed, or
distributor will require re-synchronizing.

Trim contacts as necessary. Leave the breaker arm on the pivots to
allow contacts to be checked for true mating.

# SERVICE HANDBOOK

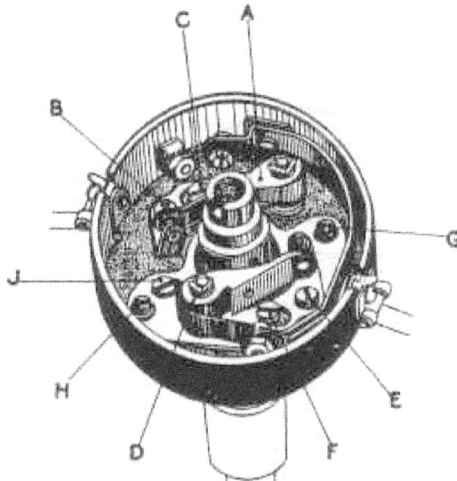

FIG. M12. DISTRIBUTOR CONTACT BREAKERS.

A. Breaker Arm.
B. Locking Screw.
(Contact plate).
C. Adjusting Screw.
(Gap adjustment).
D. Breaker Arm.
E. Locking Screw.
(Contact plate).

F. Adjusting Screw.
(Gap adjustment).
G. Locking Screw.
(Synchronizing plate).
H. Locking Screw.
(Synchronizing plate).
J. Adjusting Screw.
(Synchronizing).

Adjusting Contact Breaker Gaps:

The gaps should be set .019" to .021".      Dwell 42°

Turn engine until the fibre hub of the breaker arm(A, Fig. M12), is
on a lobe of the cam, giving maximum opening. Loosen contact plate locking
screw "B" and turn adjusting screw "C" to obtain correct gap. Repeat
operations for other arm.

Removing Distributor:

Remove distributor cover. Turn engine until the rotor arm is in
line with No. 1 cylinder firing position as indicated on the moulded cover.

Remove the two nuts securing the housing to the cylinder block,
lift off complete assembly.

DO NOT slacken the clamping plate screw, (P, Fig. M13), as the
clamping plate should be left in position; i.e., clamped to the distributor
so as not to disturb timing.

Remove the nut which secures the clamping plate to the housing "U". Remove setscrew "S" retaining the distributor to the housing.

Check cover of distributor for "tracking", indicated by a thin black line between the electrodes, in which case, a replacement must be fitted.

Automatic Governor Advance Curve:
(Silver Wraith, Silver Dawn and Bentley).

The automatic governor should conform to the following limits:-

| Advance degrees | Distributor MIN. RPM. | Distributor MAX. RPM. |
|---|---|---|
| Start | 210 | 230 |
| 1 | 220 | 250 |
| 3 | 280 | 330 |
| 6 | 440 | 500 |
| 9 | 640 | 710 |
| 12 | 870 | 960 |
| 15 | 1155 | 1255 |
| 17 | 1350 | 1450 |

The full advance should not be more than $17\frac{1}{2}$ distributor degrees.

Re-Timing Ignition and Synchronizing C.B. Arms:

1. Jack up the near rear wheel.

2. Engage 4th speed and switch on ignition.

3. With the rotor arm set on No.1 cylinder (approximately 11 o'clock), rotate rear wheel in a forward direction until timing pointer is opposite T.D.C. mark on flywheel. Check that contacts "break".

FIG. M13. DISTRIBUTOR.

K. Distributor.
L. Nut - Clamping Plate to Housing.
P. Screw- Clamping Plate.
Q. Clamping Plate.
R. Packing Washer.
S. Setscrew.
T. Driven Sleeve.
U. Distributor Housing.
V. Driven Flates.
W. Driving Shaft.
X. Vellumoid Joint.

Either of the following methods of determining precisely when the contact points separate may be used:-

(a) With the ignition switched on and a small bulb connected in series with the contact breaker points, see Fig. M14. In this case the bulb will light as the contact points "break".

(b) With the ignition switched on, observation of the ammeter will show when the points are in contact, a discharge of approximately 2 amperes will register, when the points "break", the needle returns to zero.

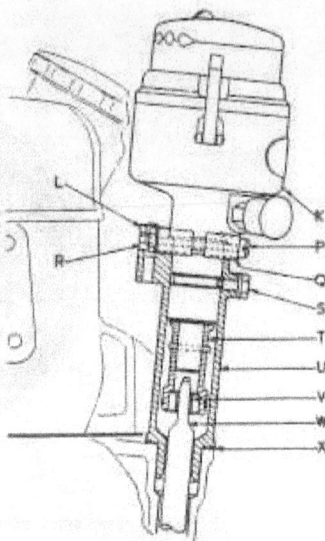

SILVER WRAITH — SILVER DAWN — BENTLEY MK. VI.

R. TYPE BENTLEY — PHANTOM IV.

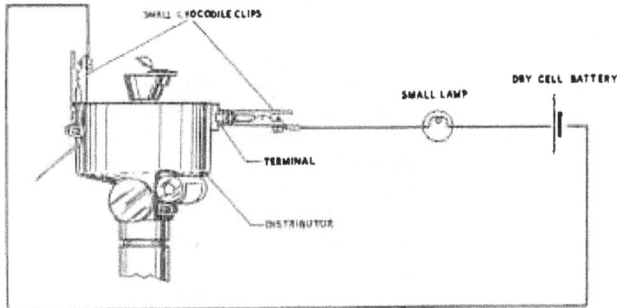

FIG. M14. CHECKING CONTACT BREAKER.

4.  Slacken screw "P"(Fig, M13), and rotate distributor head clockwise until bulb goes out, if not already so, then rotate anti-clockwise until it lights. Re-tighten screw "P", turn back rear wheel and check.

It should be noted that with premium grade fuels, the ignition may be set $2^{\circ}$ before T.D.C.

Turn the engine again until the rotor arm is set on No. 6 cylinder (approximately 5 o'clock) and set flywheel exactly as for No. 1 cylinder. Slack off screws "G" & "H", Fig. M12. Rotate screw "J" in a clockwise direction until the bulb goes out, if not already so, then rotate in anti-clockwise direction until the bulb lights. Re-tighten and check.

In the case of the Phantom IV, the contact breaker points must be synchronized with the synchronizing tool STD.410 as under.

5.  Synchronizing Phantom IV - switch on the ignition, and with the test bulb connected, place the special synchronizing tool STD.410, Fig. M15., on the cam with the "M" side of the spring in the slot of the cam.

Turn the cam clockwise until the graduations on the "M" side of the tool are near the slot in the rim of the distributor base, see Fig.M16. Continue turning until the breaker arm "A", Fig. M12., just breaks contact and note exactly the graduation on the tool that aligns with the edge of the slot.

FIG. M15. TOOL STD.410.

Continue turning until the similar graduation on the "N" side of the tool aligns with the same edge. Loosen screws "A" & "H", turn adjusting screw "J" until breaker arm "D" just breaks contact.

Check by turning cam again and re-tighten. Re-check opening of arm "D", if not between .019" to .021", re-adjust and re-synchronize.

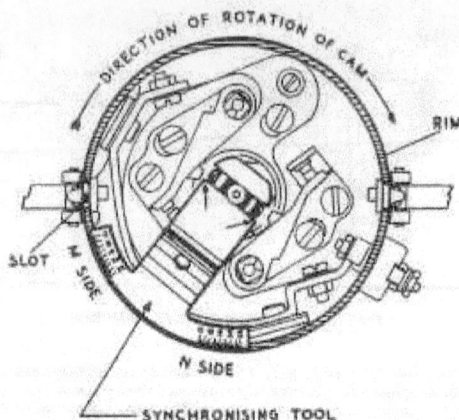

FIG. M16. SYNCHRONIZING TOOL IN POSITION.

The graduations on the tool represent engine degrees, and the markings on the "M" side are 60 distributor degrees or 120 engine degrees from the "N" side. The contact breaker arm must not vary more than 2 engine degrees.

Ignition Timing: (Automatic Gearbox Models)

The car should be run up on a ramp or over a pit. Examination of the lower bell-housing cover will show the small inspection hole on the side.

The preferable method is to note the position of the pointer and then to remove the cover. Operate the starter motor to approximately line up the flywheel marking, replace the cover for a temporary check. Remove the cover and prise the flywheel round into correct position, then permanently replace the cover.

Another method is to use the starting handle to turn the engine, which obviates the need for removing the bell-housing cover. It must be remembered that the starting handle operates through the friction damped spring drive unit, and therefore, allowance must be made for the wind-up of the spring drive unit. If it is decided to use this method, the timing should be set .300" late of the IGN.TDC mark on the periphery of the flywheel.

The contact breaker should now be adjusted by rotating in an anti-clockwise direction, so that the cam is just on the point of causing the contact break when revolving in the normal direction, while at the same time the high-tension rotor is opposite No. 1 distributor contact, the rotor being in the fully retarded position.

IGNITION COIL:

The ignition coil is wound for the positive earth system.

The coil terminals are marked "CB" for contact breaker and "SW" for switch lead. Connect the lead from the radio suppressor to the "SW" terminal.

The outside of the coil casing should be kept clean; misfiring is occasionally caused by an accumulation of dirt around the terminals.

SPARKING PLUGS:

The sparking plugs for the Silver Wraith, Silver Dawn and Bentley may be either Champion Type N8 or Lodge Type CLN, 14 m/m.

The gaps for the above plugs should be .025" where no television suppressor is incorporated in the high tension lead to the distributor, and .030" when this suppressor is fitted.

The sparking plugs for the Bentley Continental Sports Saloon are Champion NA.8 only.

The sparking plugs for the Phantom IV are Champion N8 only.

HEADLAMPS:

On the Silver Wraith and Phantom IV the headlamps are Lucas Type R.100.

On the Silver Dawn and Bentley the headlamps are the Lucas "Built-in" Type either Mark I or Mark II. These lamps incorporate a Light Unit, which consists essentially of a reflector and front glass assembly provided with a mounting flange, by means of which it is secured in the body housing. Normally, the pre-focus flange type bulb is fitted. For France, an adaptor and 3-pinned bulb is fitted, and for North America, the regulation "Sealed Beam" unit is fitted. The bulbs are correctly positioned in relation to the focal point of the reflector, and it is not necessary to re-focus when a new bulb is fitted.

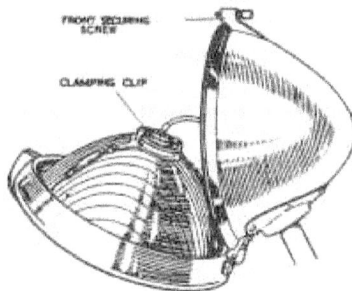

FIG. M17. HEADLAMP R.100.

Mark II headlamps incorporate a fuse unit in the lamp shell, these fuses are rated at 15 amperes.

Changing Bulb, R.100 Type:

The lamp front and reflector can be swung downwards, Fig. M17, if the screw at the top is slackened off. The bulb holder can then be removed from the base of the reflector.

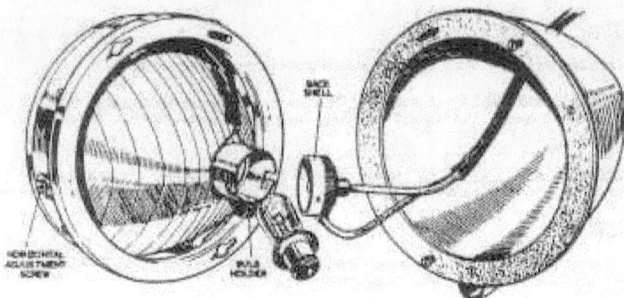

FIG. M18. HEADLAMPS, MK.I, CHANGING BULB.

## Changing Bulb, Mk.I Type:

Slacken the screw at the bottom of the lamp, lift off the rim, removing it from the bottom first. Slacken the screws securing the flange of the Light Unit and turn unit anti-clockwise to detach.

Twist back-shell of bulb holder in an anti-clockwise direction and remove and extract bulb.

## Changing Bulb, MK.II Type:

Slacken off screw at bottom of outer rim and remove rim and dust excluder.

Push Light Unit inwards against the spring-loaded screws, turn unit anti-clockwise and withdraw. Remove bulb as for Mk.I.

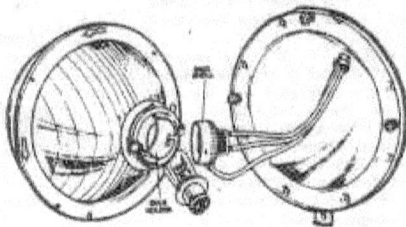

FIG. M19. HEADLAMPS, MK.II, CHANGING BULB.

## Setting and Focussing:

When setting the lamps, the measurements between ground level and lamp centres are made with the car loaded with five persons.

The headlamps must be set so that the beams are directed straight ahead and are parallel with the ground and with each other.

R.100 Type - To adjust, slacken fixing nut at the base of the lamp and move the lamp on its adjustable mounting as required.

Cover up one lamp while testing the other. If the lamp gives a uniform long-range beam without a dark centre, the bulb needs no adjustment. The bulb holder can be moved backwards or forwards for focussing when the clip at the back of the reflector is slackened. After adjustment, retest with reflector and front fitted.

SILVER WRAITH — SILVER DAWN — BENTLEY MK. VI.

R. TYPE BENTLEY — PHANTOM IV.

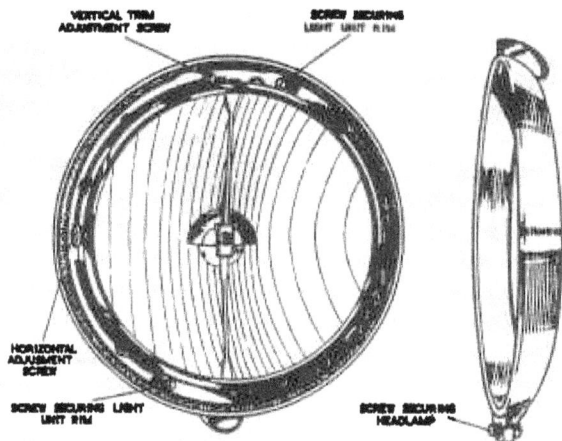

FIG. M20. HEADLAMP, Mk.I, FOCUSSING.

Mk.I Type - To adjust, remove front rim. If vertical adjustment is required, set with vertical true adjustment screw at top of reflector unit. Turn screw clockwise to raise the beam and anti-clockwise to lower.

If horizontal adjustment is necessary, slacken off the two horizontal adjustment screws, one each side of Light Unit. The reflector may then be positioned as required.

Mk.II Type - To adjust, remove front rim. Adjustment is by setting one of the three spring-loaded screws which retain the Light Unit. The top screw is for vertical beam trimming and the side screws for trimming horizontally.

PASS LAMPS:

The Silver Wraith and Bentley pass-lamps are similar, as illustrated in Fig. M22. They are fitted with a removable light unit, the rim being screwed at the bottom with either a spring clip or screw.

The Phantom IV and Silver Dawn pass-lamps are similar, as illustrated in Fig. M23. They are fitted with a removeable light unit; the rim forming a clip, secured at the bottom by a screw.

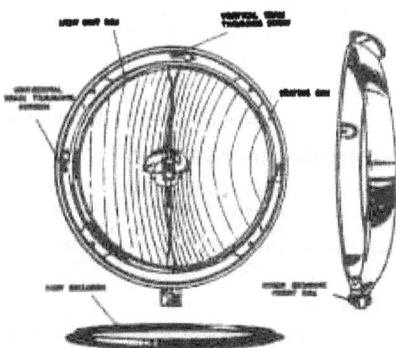

FIG. M21. HEADLAMP, MK.II, FOCUSSING.

- M.17 -

SILVER WRAITH — SILVER DAWN — BENTLEY MK. VI.

R. TYPE BENTLEY — PHANTOM IV.

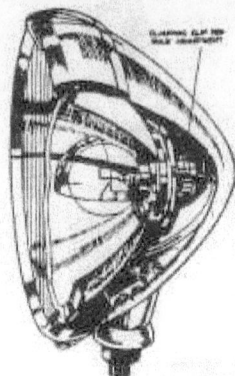

FIG. M22. PASS-LAMP.

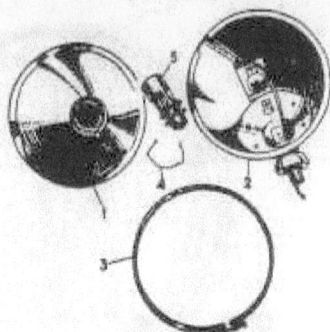

FIG. M23. PASS-LAMP - DISMANTLED.

| | |
|---|---|
| 1. Light Unit. | 4. Spring ring. |
| 2. Lamp bowl. | 5. Bulb. |
| 3. Rim. | |

To change the bulb, remove light unit, the bulb holder can then be removed from back of reflector.

When setting the lamps, set the nearside lamp or centre lamp .250" off centre to allow for camber of road.

SIDE LAMPS:

To renew the bulb, remove the securing screw from the top and withdraw the assembly. To obtain access to bulb, detach the front portion by holding firmly and twisting the rear portion a quarter of a turn to release the spring catch.

STOP TAIL LAMP:

To renew bulb, remove small screw and withdraw front rim and glass.

NUMBER PLATE AND REVERSE LIGHTS:

Three types have been fitted, inspection will show that to remove the bulb either remove securing screw to remove glass and rim, or where no screw is visible, unscrew glass and rim to expose bulbs. On cars fitted with the square number plate, remove fibre cover in boot lid by pulling off spring clips to expose nut and bolts securing lamp to boot cover.

SWITCHBOX:

The switchbox incorporates the Master Switch, which controls all the electrical system except clock, roof lamp and inspection lamp; the ignition switch, the starter button, and a Yale lock.

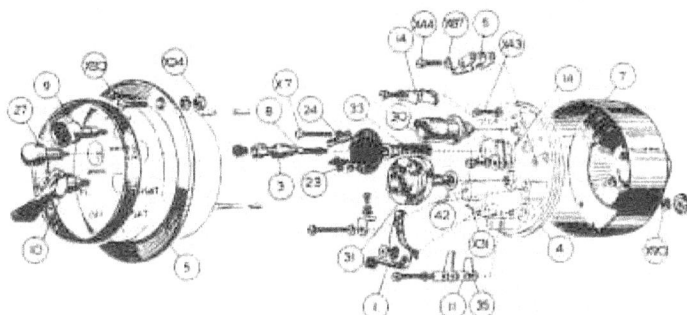

FIG. M24. SWITCHBOX - DISMANTLED.

| | | |
|---|---|---|
| 1. | Main contact arm. | 11. Packing. |
| 3. | Brush, starter contact. | 14. Packing. |
| 4. | Switch base. | 23. Backing plate. |
| 5. | Body. | 24. Contact plate. |
| 6. | Brush accessories. | 27. Starter push. |
| 7. | Cover. | 30. Spindle. |
| 8. | Carbon contact. | 31. Spindle, main. |
| 9. | Operating lever. | 33. Spring. |
| 10. | Operating lever. | 35. Stop, brush. |
| | | 42. Thrust washer. |

To Examine and Test Switchbox:

Remove switch from facia board and then remove base plate. Note, do not dismantle further than is necessary to replace worn or defective parts.

(i)     Check that the Master Switch and Ignition Switch turn freely without end play or radial slackness. The movement from one position to another, controlled by the cams operating on spring loaded balls, should be definite but without harshness or appreciable slackness.

(ii)    The two solid contacts to terminals 4 and 5, should be free on thin pivots and each have a pressure where in contact of not less than 14 ozs.

(iii)   Check the bedding of the four fixed contacts of the Starter Switch on the moving contact.

Grease all contacts with vaseline and re-assemble switch, check electrical operation by connecting negative terminal to a 12 volt battery, and with a 12 volt lamp in series with the battery positive.

(i)     With Master Switch in OFF position and the Ignition and Starter Switches ON, no light should be obtained by connecting lead to any terminal.

(ii)   With the Master Switch at P.L. and the other switches on, light
       should be obtained at No. 5 terminal only.

(iii)  With the Master Switch only in the ON position, light should be
       obtained from No. 8 terminal only.

(iv)   With Master Switch and Ignition ON, light should be obtained from
       terminals 2, 6, 8 and 11.

(v)    With Master Switch in S & T position, light should be obtained from
       terminals 2, 5, 6 and 8. Now link terminals 8 and 2 together and
       light should be obtained from terminals 2, 6  and 8.

(vi)   With the Ignition OFF and starter button depressed, light should
       be obtained from terminals 2, 6, 8 and 10.

(vii)  With the Master Switch at H. S & T, light should be obtained from
       terminals 2, 4, 5, 6 and 8.

<u>HORNS</u>:

          The horns are of Lucas
manufacture.

<u>Testing the Relay</u>:

          Operate the horn push and note
if the relay armature moves and the
contacts close. If not, there is a
fault in the relay itself, or in the
wiring. Remove the horn push leads
from the relay unit and connect a 12
volt supply across the terminals, if
the relay is in order, the contacts
will close. It is set to operate at
7 - 8 volts.

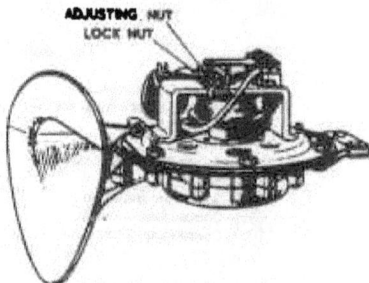

FIG. M25.  HORN.

<u>Adjustment</u>:

          If the contacts are dirty, clean with glass-paper.

          When adjusting, disconnect the supply lead from the
other horn, also remove the fuse as it is likely to blow.

          Adjustment does not alter the pitch of a horn note,
but takes up wear of moving parts.

<u>TRAFFICATORS</u>:

          The trafficators are of Lucas manufacture.

<u>Servicing</u>:

          If the arm is sluggish in rising, apply a drop of
thin oil to the catch pin between the arm and operating
mechanism, see Fig. M26.

FIG. M26.  LUBRICATING
           CATCH PIN.

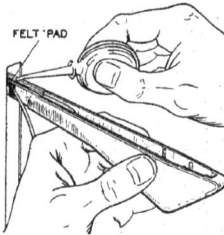

FIG. M27.  LUBRICATING
          FELT PAD.

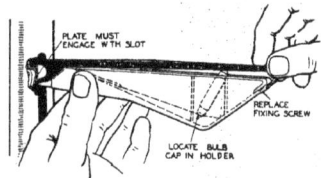

FIG. M28.  REPLACING COVER PLATE.

To oil the felt pad in the top of the arm, lubricate the spindle bearing, see Fig. M27.  Withdraw the screw on the underside of the arm, slide off the metal cover.

Failure to light up, usually indicates a bulb failure or imperfect contact.

## DIRECTION INDICATOR LIGHTS:

The winking light type of direction indicator is fitted to certain export models, and is operated by a flasher unit mounted on the valance plate.

When the signal system is operating properly, the lights flash about 90 times per minute.  If either front or rear signal bulb is burned out, the reduced current will increase the flasher speed and the pilot bulb on the facia board will not light.

A clicking noise in the flasher unit makes an audible signal when the circuit is on, and this is purposely created as an additional warning that the unit is operating correctly.

The signal flasher is a sealed unit and is non-adjustable, and if service is necessary it must be by replacement.

## WINDSCREEN WIPERS:

Early cars were fitted with the Houdaille Co's. "Berkshire" wiper. Later, the Lucas single speed wiper was fitted, on current production, the Lucas 2 speed wiper is standard.

It should be noted that the 2 speed wiper motor incorporates a thermostatically controlled cut-out which cuts off the current supply if the wiper is overloaded.

Complaints of wiper motor failure may be due to the normal operation of the cut-out, which may operate if the wiper is run at high speed on dry or partly dry screen.

The thermostat will reset itself if the wiper is switched off and left for 10 - 15 minutes.

ELECTRICAL WIRING DIAGRAM.
SILVER DAWN CHASSIS SERIES SBA TO SHD. (R.H. DRIVE)

VOLTAGE REGULATOR
IGNITION DISTRIBUTOR
6 5 4 3 2 1
IGNITION COIL
DYNAMO
COACHBUILDERS JUNCTION BOX
TRAFFICATOR SWITCH
CHARGING PLUG
SWITCH-BOX
MAP LAMP
HEATER RHEOSTAT SWITCH
DE-MISTER SWITCH
FOG LAMP SWITCH
PETROL TANK UNIT
OIL SUMP UNIT
HORN RELAY UNIT
R.H. SIDE LAMP
R.H. HEAD LAMP
R.H. FOG LAMP
TO H.T. PUSH IN COLUMN END
R.H. FRONT JUNCTION BOX
IGNITION JUNCTION BOX
OIL GAUGE & THERMOMETER
CLOCK
IGNITION WARNING LIGHT
MAP LAMP SWITCH
FUEL OIL/LEVEL SWITCH
R.H. HORN
L.H. HORN
SPEEDOMETER
L.H. FOG LAMP
AMMETER
DISTRIBUTION BOARD
L.H. HEAD LAMP
FOOT SWITCH
FUEL WARNING LIGHT
CIGAR LIGHTER
L.H. SIDE LAMP
R.H. STOP & TAIL LAMP
L.H. FRONT JUNCTION BOX
REVERSE LAMPS
NUMBER PLATE
DE-MISTER BLOWER
BODY JUNCTION BOX
WIPER MOTOR
FUEL LEVEL INDICATOR
INSTRUMENT LIGHTS SWITCH
STARTER MOTOR RELAY SWITCH
STARTER MOTOR
SCREEN WIPER SWITCH
RADIO
TAIL JUNCTION BOX
BOOT LAMP SWITCH
STOP LAMP SWITCH
LUGGAGE BOOT LAMP
REVERSE LAMP SWITCH
PETROL PUMPS
SINGLE BOX CONTROL SWITCH
BATTERY
R.H. STOP & TAIL LAMP
P.L.
OFF
ON
SAT
H.SAT
DIAGRAM OF MASTER SWITCH

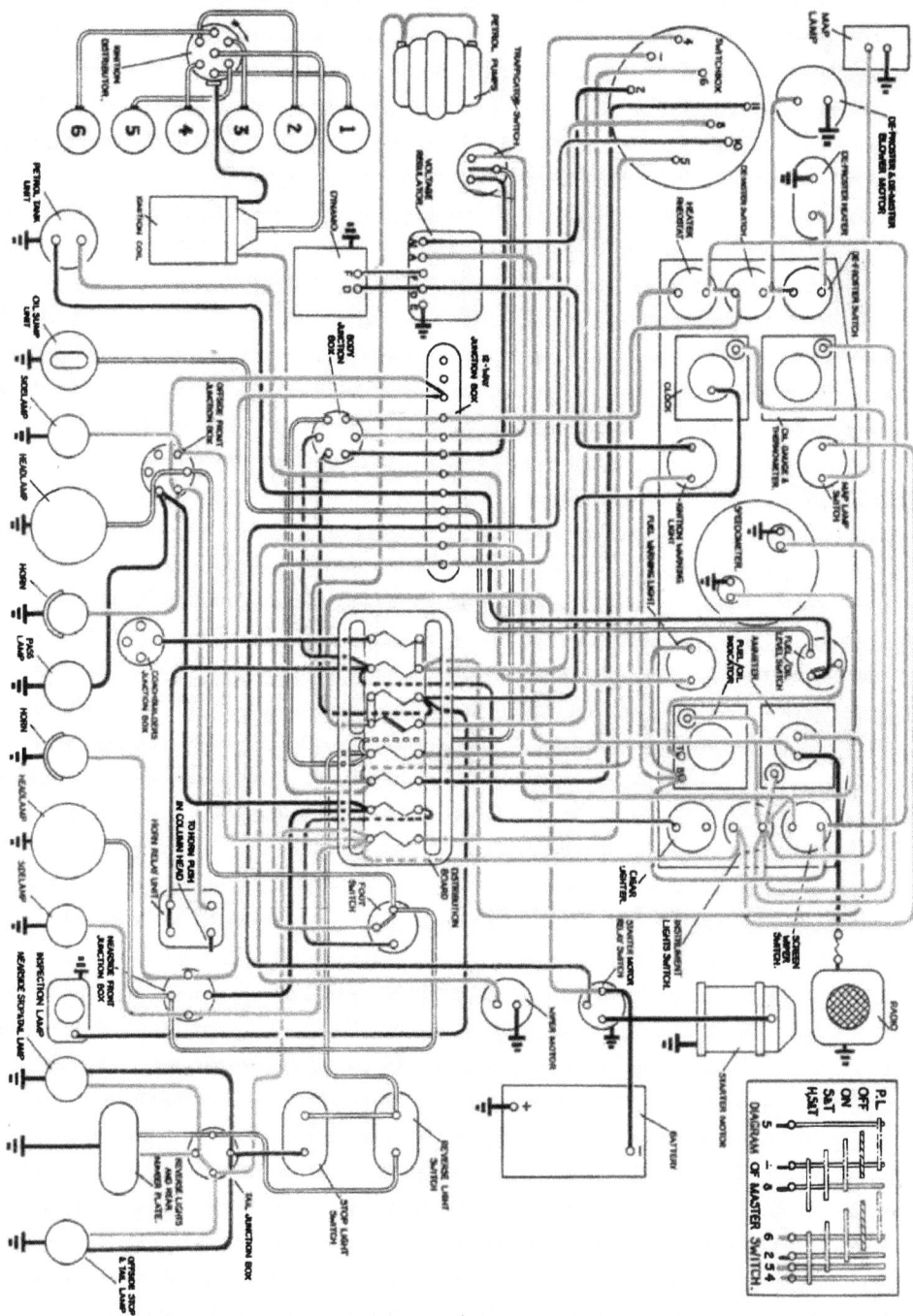
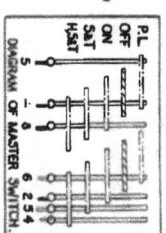

ELECTRICAL WIRING DIAGRAM.
SILVER WRAITH CHASSIS SERIES WGC TO WSG (R.H. DRIVE)

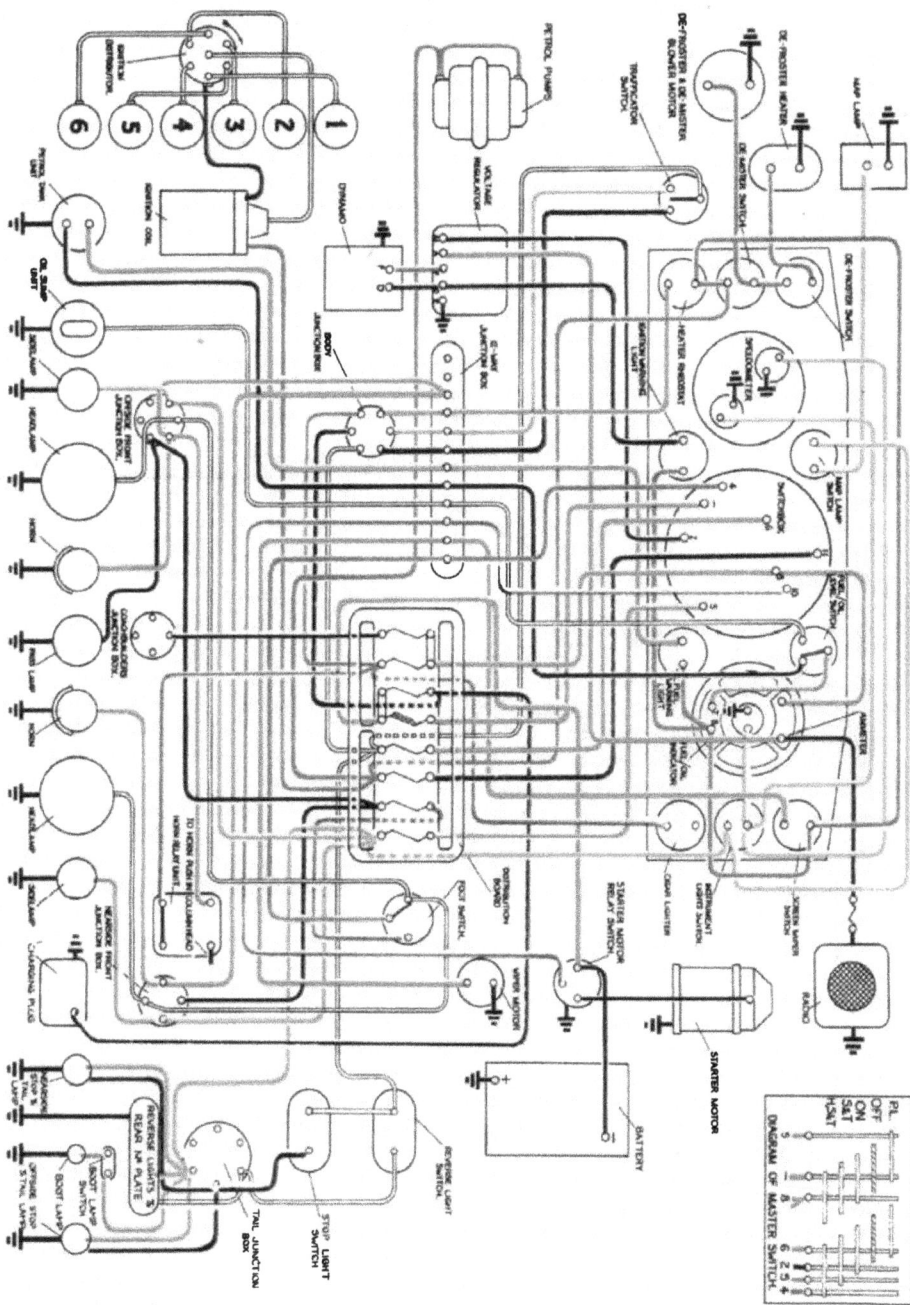

ELECTRICAL WIRING DIAGRAM

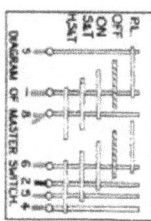

DIAGRAM OF MASTER SWITCH

ELECTRICAL WIRING DIAGRAM.
BENTLEY CHASSIS SERIES DZ TO PU. (R.H. DRIVE)

ELECTRICAL WIRING DIAGRAM.
BENTLEY CHASSIS SERIES "R" TYPE. (R.H. DRIVE, NON-AUTOMATIC)